Air Forces of Latin America

Colombia

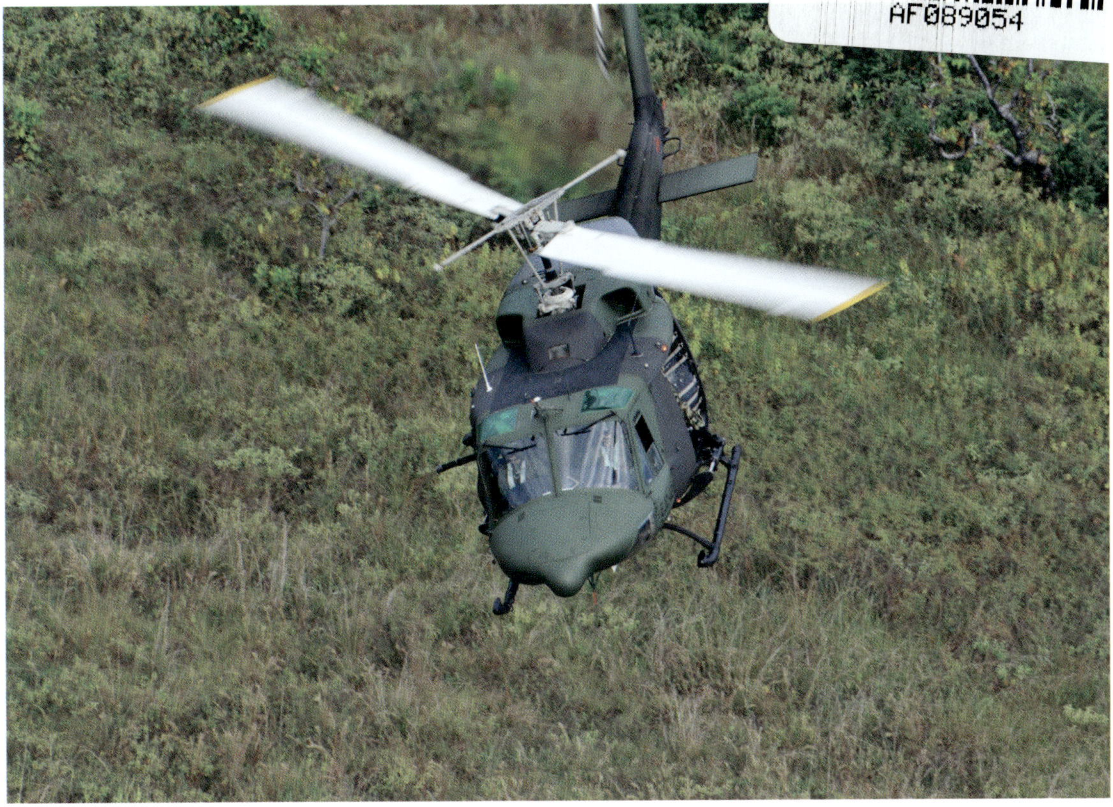

SANTIAGO RIVAS

Key Books

AIR FORCES SERIES, VOLUME 5

Front cover page: An IAI Kfir C10 of CACOM-1 of the Fuerza Aérea Colombiana (Colombian Air Force) showing part of the weapons that the type has in its inventory.

Back cover page: An AC-47 'Fantasma' prepares to take off on a night mission at Palanquero, while another is standing behind.

Title page: A CUH-1N, armed with a GAU-17 minigun and equipped with extra fuel tanks.

Contents page: Plans for the future include the purchase of two extra CN-235-300MP Persuaders, to increase the maritime patrol capacity. The three currently operational aircraft are very active in the fight against drug traffickers.

Published by Key Books
An imprint of Key Publishing Ltd
PO Box 100
Stamford
Lincs PE19 1XQ

www.keypublishing.com

The rights of Santiago Rivas to be identified as the author of this book has been asserted in accordance with the Copyright, Designs and Patents Act 1988 Sections 77 and 78.

Copyright © Santiago Rivas, 2022

ISBN 978 1 80282 196 3

All rights reserved. Reproduction in whole or in part in any form whatsoever or by any means is strictly prohibited without the prior permission of the Publisher.

Typeset by SJmagic DESIGN SERVICES, India.

Contents

Introduction 4
Chapter 1   Fuerza Aérea Colombiana – Colombian Air Force 5
Chapter 2   Aviación Naval Colombiana – Colombian Naval Aviation 43
Chapter 3   Aviación del Ejército Colombiano – Colombian Army Aviation 56
Chapter 4   Policía Nacional de Colombia – Colombian National Police 77
Chapter 5   Orders of Battle 89
Glossary 95

Introduction

Colombia has the Ministerio de Defensa Nacional (Ministry of National Defence), under which are the Ejército Nacional de Colombia (EJC, Colombian National Army), Armada Nacional de la República de Colombia (ARC, Colombian National Navy), Fuerza Aérea Colombiana (FAC, Colombian Air Force) and Policía Nacional de Colombia (PNC, National Police of Colombia), the last being the only security force in the country. Each of these forces has its own aviation division and, due to the internal situation of the country, they are all very powerful, with the biggest combined aviation force in the world.

Aviation within the Fuerzas Militares de Colombia (Military Forces of Colombia) had a slow start. The first attempts were made in 1916, but it was not until 1920 that the first air operations began. The crisis with Peru over the Putumayo area in 1932 and 1933 was the real start of military aviation and also its consolidation. During and after World War Two, Colombian military aviation was modernised, and, in 1949, it began its long history of counterinsurgency operations. The need for these operations came after the killing of politician Jorge Eliecer Gaitán led to the emergence of insurgency groups and a period called 'La Violencia' ('The Violence'), which lasted for the entirety of the 1950s and later evolved into bigger guerrilla groups and gangs of drug traffickers; a problem that still affects the country today.

Although the ARC was operating a few aircraft in the mid-1930s, until the late 1950s, the FAC was the only military and security force to operate aircraft. Then, the PNC started to operate a few aircraft, while, in 1968, a group of civilians and ARC officers organised a small naval air division with light aircraft, which, in 1983, became the Comando de Aviación Naval (Naval Aviation Command). The EJC was the last force to have its own aviation division, which began formally in 1997. The intensity of the internal conflict and the drug production in the 1980s led to a fast growth of the aviation units of all forces, not only in terms of equipment, but also in personnel and training.

In addition to internal conflict, the increasing tension with Venezuela, as well as the intention of Colombia to become a relevant actor in the wider Latin American defence network, has meant that, in the past ten years, the three Fuerzas Militares de Colombia have introduced new equipment and doctrine for conventional war, with an increase in the air-to-air and anti-tank capabilities, among others, while the Comando de Aviación Naval is now looking for anti-ship and anti-submarine capabilities. Participation in exercises inside and outside Colombia has also increased considerably, including in *Red Flag*, *Maple Flag*, *Cruzex* and many others.

Although, in 2015 and 2016, peace agreements with the FARC (Fuerzas Armadas Revolucionarias de Colombia, Revolutionary Armed Forces of Colombia) terrorist group were negotiated, in an attempt to bring the internal war to a close, they failed to put an end to the conflict. Many factions of the FARC, as well as the other main terrorist group, the ELN (Ejército de Liberación Nacional, National Liberation Army), continue to operate and, thanks to the reduction of actions against them and the growing support from Venezuela, have gained more power.

While plans for new fighters, medium helicopters and other types of aircraft have been developed in recent years, the economic crisis of the country, beginning in 2019 and worsened by the COVID-19 pandemic, led to a reduction in the defence budget, and most of the programmes are now progressing very slowly.

Chapter 1
Fuerza Aérea Colombiana – Colombian Air Force

While the first (failed) attempt to create a division of military aviation in Colombia took place on 17 September 1916, the Escuela Militar de Aviación (School of Military Aviation) was officially created on 31 December 1919 at Flandes, 100km to the southwest of Bogotá, becoming the Fifth Arm of the EJC, and later the Sección de Aviación Militar (Military Aviation Section) of the Ministerio de Guerra (War Ministry). However, its life was very short, and it was closed in 1922. In 1924, however, it was reopened at Base Arsenal Madrid, close to Bogotá, with a few Wild WT34D and Comte Wild-X aircraft. Growth was very slow until September 1932, when a conflict erupted with Peru, and the latter occupied the town of Leticia in the Amazon. The force received the support of the local airline SCADTA, from which it bought several aircraft, including Junkers F 13s and W 33s, a Dornier Wal and a Merkur, a Fokker Super Universal and an aerodrome at Palanquero, on the shores of the Magdalena River. In December of 1932, the División General de Aviación Militar (General Military Aviation Division) was founded as part of the Departamento 8 (Department 8) of the War Ministry, and more aircraft were bought directly from the factories, including three Junkers K 43s, seven W 34s and six Ju 52/3ms, 30 Curtiss Model 35A Hawk II fighters, two Consolidated P2Y-1C flying boats and 25 Curtiss Model 37F Cyclone Falcons. The conflict with Peru, which ended in May 1933, showed the importance of military aviation, and other bases were created and new aircraft acquired, including five Curtiss Wright CW-16s, 20 Consolidated PT-11C Huskys, five Fairchild 22-C7Fs, four Ford 5-AT-C Trimotors, three Curtis BT-32 Condors, four Bellanca Air Bomber 77-140s and four Seversky SEV-3M-WW equipped with floats. On 15 July 1942, the División General de Aviación Militar became the Fuerza Aérea Nacional, but was still formally part of the EJC, until, on 31 December 1944, the Fuerza Aérea Colombiana was born as an independent force.

During World War Two, despite Colombia giving its support to the Allies, they only received training aircraft in return, including 30 Boeing Stearman PT-17 Kaydets in 1942 (followed later by another 30), 100 North American AT-6B, C and D Texans (the first six had arrived in 1940), 20 Vultee BT-15 Valiants from 1943, and, from 1944, a large fleet of Douglas DC-3/C-47s, which grew over the years. After the war, the FAC began to receive new aircraft through the American Republics Project of the United States, including, in 1946, 35 Republic P-47D Thunderbolt fighters, 13 Consolidated PBY-5 Catalinas and four North American B-25J Mitchells, with which the FAC modernised its combat force. In 1953, the first helicopters arrived, two Hiller H-12As and four OH-23Bs. In 1954, to replace the basic trainers, they purchased 50 Beechcraft T-34A Mentors, followed later in the 1970s by another ten, plus six T-34Bs. In 1954, the FAC entered into the jet age with the first six Lockheed T-33 Silver Stars from a total of 52 that it would receive over the years, the last 12 arriving in 1978. These were followed from 1955 by 15 Lockheed F-80C Shooting Stars (with ten more in 1963), and, in the same year, the attack force was reinforced with 20 Douglas B-26C Invaders, while six Canadair Sabre Mk.IVs arrived in 1956, followed by four North American F-86F aircraft in 1963.

While the internal situation in the country has been tumultuous since the late 1940s, it became more complicated in the 1960s, with the appearance of different guerrilla groups and the creation, in 1964, of the ELN, and, in 1965, of the FARC, the two main terrorist groups that still exist today. These circumstances led to a change in the focus of the Fuerzas Militares de Colombia from traditional warfare to counterinsurgency operations, which in turn meant an increase in the number of helicopters operated by the FAC; the first ten Bell UH-1Bs arrived in 1963 and the first 40 UH-1H in 1969, together with 15 Hughes 369HS. In 1968, other acquisitions included the first 15 Cessna T-37C Tweets and two Lockheed C-130B Herculeses.

In 1959, the Comando de Bombardeo (Bombardment Command) was created at Base Aérea (BA) Apiay, to the east of Bogotá, and, in 1966, the Comando Aéreo de Combate (Air Combat Command) was organised at BA Palanquero. In 1971, these became the Comando Aéreo de Combate No 1 (CACOM-1) at Palanquero and No 2 (CACOM-2) at Apiay; these commands remain in these locations today, however, there are now seven CACOMs in total.

In 1972, the fighter division of the FAC took a big step forward with the purchase of 14 Dassault Mirage 5COAs, two 5CODs and two 5CORs (R for reconnaissance); 34 Cessna A-37B Dragonflies arrived in 1977 to aid counterinsurgency missions; and, in 1988, the first 14 IAI Kfirs were received to increase the FAC's attack power after a crisis with Venezuela. After receiving the first of a large force of Bell 212s and CUH-1Ns in 1971, in 1988 the force started to add the Sikorsky UH-60 Black Hawk. The counterinsurgency fixed-wing fleet also grew, with three FMA IA-58A Pucarás in 1989 and 22 North American Rockwell OV-10A Broncos and 14 Embraer EMB-312 Tucanos in 1991, which were joined in 2006 by 25 EMB-314 Super Tucanos.

Owing to its continuing internal conflicts, the force is one of the most experienced of its kind in Latin America when it comes to counterinsurgency measures, with almost 60 years of operations under its belt. However, the threat posed by Venezuela, whose government openly supports the guerrillas and drug traffickers and has threatened Colombia on many occasions, has, in the past decade, led to a shift in the FAC's focus away from counterinsurgency and back once again to conventional operations. In support of Venezuela, the violation of the FAC by Russian ELINT (electronic intelligence) Ilyushin Il-96-400VPUs became commonplace in the skies over the Caribbean coasts of the country, with the Kfirs managing to intercept them several times, the last time being 19 April 2021.

While the vital Kfir fleet was reinforced and modernised in 2009, the Mirages were retired in 2010, and the FAC is now looking for their replacements. It initially selected the Lockheed Martin F-16 Block 70, after a selection process that also included the Eurofighter Typhoon, the F/A-18E Super Hornet, the Dassault Rafale and the Gripen NG, but the government is yet (as of February 2022) to authorise this purchase.

The development of the Sikorsky AH-60L Arpía IV, the latest version of the armed Black Hawks, included the addition of different versions of Spike anti-tank missiles, with the aim of countering the threat of Venezuelan tanks, which are a big threat in the desert areas of La Guajira, in the northeast of Colombia, as the EJC has no tanks. Furthermore, the FAC has just replaced its last T-37Bs with six Beechcraft T-6C Texan IIs, which will be followed by another batch of six, with the intention of reaching a total of 24 examples.

Current organisation
Dependant to the head of the force, Comando de la Fuerza Aérea (Air Force Command), are the Comando de Operaciones Aéreas y Espaciales (Air and Space Operations Command), which is in

charge of conducting and ordering the operations of the FAC; the Comando de Apoyo a la Fuerza (Force Support Command), which is in charge of logistics, security and communications; the Comando de Desarrollo Humano (Human Development Command), which is in charge of everything related to personnel; seven CACOMs; the Comando Aéreo de Transporte (CATAM, Air Transport Command), which has medium and large transport aircraft, in addition to carrying out the VIP transport of the forces and the government; and the Comando Aéreo de Mantenimiento (CAMAN, Maintenance Air Command), which oversees the major maintenance of the aircraft. There are also three education centres – the Escuela Militar de Aviación, Escuela de Posgrado (Postgraduates School) and the Escuela de Suboficiales (NCOs School) – and four air groups – Grupos Aéreos del Oriente, Caribe, Casanare and Amazonas (Air Groups of the East, Caribbean, Casanare, and Amazon). Finally, there is the Fuerza de Tareas Conjunta (Joint Task Force) Ares, used for operations against drug traffickers, and the Grupo de Operaciones Especiales Aéreas (GROEA, Air Special Operations Group) of special forces.

The CACOMS

CACOMs carry out offensive and defensive missions, as well as providing tactical support for them, having more than one squadron each. Each squadron has one or more escadrille under its command. Air groups, for the most part, perform the same tasks but are smaller units with only one squadron. Each CACOM has its own base, with ground security personnel and workshops to perform most stages of maintenance, except major checks, which are performed at CAMAN. In the cases of CACOM-1 and 2, they have their own firing range inside the base. In terms of personnel, the FAC has about 1,000 pilots, 2,300 officers, 3,300 NCOs, of which 1,900 are aeronautical technicians, 463 students, 495 cadets, 4,600 soldiers and 2,500 civilian employees.

CACOM-1

This is the oldest unit of the FAC, after the Escuela Militar de Aviación, having inherited the airfield at La Dorada, on the shore of the Magdalena River, built by SCADTA in the 1920s, which was transferred to the military aviation in 1932 during the Putumayo conflict with Peru. Currently, the base is called Germán Olano, but, since the place is called Palanquero, it is also known by that name. Since then, it has been the main base aérea of the FAC and the home of fighter aviation in Colombia, and, until 1959, it was the only combat base of the force. Its air unit comprises the Grupo de Combate 11, which in turn comprises four squadrons.

The IAI Kfir is currently the main fighter aircraft of the Fuerza Aérea Colombiana (FAC, Colombian Air Force). However, there are plans to replace them, most probably with F-16 Fighting Falcons.

Escuadrón de Combate 111 'Dardos' (Combat Squadron 111 'Darts')

This squadron is the main fighter unit of the FAC and has the entire fleet of IAI Kfir C10, C12 (known in Colombia as 'COA') and TC12 ('COD') aircraft. While Colombia wanted to buy Kfirs to reinforce their fleet of Mirage 5s in 1981, the US vetoed the operation until 1987, and, finally, in 1988, Colombia finally purchased a batch of 12 second-hand IAI Kfir C2s from Israel, modernised to C7 standard, and one Kfir TC2 upgraded to TC7 standard. The aircraft started to arrive in Colombia on 28 April 1989 and formed the Escuadrón 213 in the 1990s, starting operations against guerrilla groups shortly after.

To increase the efficiency of the missions, a new upgrade programme was launched in February 2001. The single-seaters received new night-vision systems, a WDNS-391 weapons delivery and navigation system for smart weapons delivery, an Elbit 82 stores management system, video capability and an armament control display panel. The two-seaters received a Cockpit Laser Designator System (CLDS) to direct the new IAI Griffin laser-guided bombs, which were purchased in the same year as the upgrade programme began. Shortly after, they also received Rafael Python III air-to-air missiles.

By the beginning of 2008, the FAC had announced the signing of a contract with IAI for the purchase of 11 Kfir C7s and three TC7 aircraft to replace the ten remaining Mirage 5s (one C7 was used for

Above: The Kfirs were extensively modernised, gaining new avionics, an Active Electronically Scanned Array (AESA) radar and an Electronic Counter Measures (ECM) suite.

Left: Among the weapons used by the Colombian Kfirs are the Phyton 5 and Derby R missiles and Paveway II guided bombs.

Above: A Kfir flies together with the Boeing KC-767 tanker aircraft. The KC-767, baptised 'Jupiter', replaced a KC-707.

Right: Two Airbus CN-235s were modified into the ECN-235-100M Phobos standard, but one was lost in an accident in 2015. The other is currently the primary electronic warfare aircraft in Colombia.

spares). The contract included an upgrade of all these aircraft, plus the ten C7s and one TC7 already in FAC service. Seven of the new aircraft were taken to C10 standard and the other three to C12, while the two-seaters were taken to TC12 standard. Of the 13 pre-existing C7s, seven were taken to C12 and three to C10. The C10s were equipped with the Elta EL/M-2032 radar, while the C12 had the Elta EL/M-2001B telemetry radar. All of the aircraft received new avionics, liquid crystal displays (LCDs), Elbit DASH helmets, Link 16 datalink systems, one-piece windshields and the capacity to fire Python 5 and Derby missiles, which were also acquired, considerably increasing the combat capabilities of the aircraft. They also received Litening and RecceLite pods to add to the arsenal of Griffin bombs.

The first example was handed over in Israel in April 2009. Soon after, deliveries began in Colombia. One of the new two-seaters, FAC 3004, was lost on 20 July 2009 when taking off from Cartagena for a pre-delivery test flight with Israeli pilots and was replaced by another, FAC 3007, while the other two-seater was lost shortly after entering service. In September 2009, flight tests of the first modified Kfir C10 began.

After the new Kfirs were received, in December 2010, the Mirage 5COAMs and CODMs were retired from service, leaving the Kfirs as the only jet fighters in service in Colombia. Initially, the older aircraft served with Escuadrón 111 and the new ones at Escuadrón 112 (both units changed the numbers in the 1990s), but shortly after all went to serve with the 111. The high degree of training undertaken by the pilots of these aircraft was recognised with an invitation to take part in the 2012 edition of the *Red Flag* exercise in the US, and on 11 July 2012, seven Kfir C10s and one TC12 arrived at Nellis Air Force Base, ready to participate.

By 2015, all but one of the C12s were taken to C10 standard, leaving 18 C10s in service, now called Kfir COA, and the El/M-2032 radar was replaced by the El/M-2052 AESA radar. They also received the Elbit Systems Emerald AES-212 electronic warfare system. With the aircraft receiving the Elbit Spectrolite SPS-65V6 jamming system, some received the designation 'C10 EW' on the nose, to the left of the Kfir COA badge. In April 2015, two extra TC12s were delivered to replace two lost to accidents and keep a fleet of three two-seaters.

To replace the Griffin guiding system, new GBU-49 Paveway II kits for laser/GPS guidance were purchased, as well as Rafael Spice electro-optical/GPS guidance system for Mk.83 bombs. Around 2018, the FAC purchased new kits for Mk.82 bombs, but these will be mainly used on Super Tucanos. Furthermore, I-Derby ER missiles, with a range of up to 100km, were added to the inventory, while Targo systems replaced the DASH ones on the helmets.

After the successful participation in *Red Flag* in 2012, on 17 July 2018, six Kfirs and a Boeing KC-767 returned to Nellis to take part in the exercise from 23 July to 3 August.

As previously mentioned, the FAC is currently looking for a replacement for its Kfirs, and they have chosen the F-16 Block 70, with the intention of buying 15 aircraft with ten options, with the aim of reaching a total of 48 aircraft by 2040. These should be equipped with an AN/APG-83 Scalable Agile Beam Radar (SABR), General Electric engines, and an undisclosed weapons package of US$320m was discussed in the contract. While the FAC was expected to sign a contract in 2020, with the intention of having the first aircraft in 2023, the economic crisis in the country has delayed those plans, and the future of the programme is now uncertain.

Escuadrón de Combate Táctico 113 'Fantasma' (Tactical Combat Squadron 113 'Ghost')

This unit is the primary operator of the Basler AC-47T 'Fantasma' gunships, having two aircraft on strength from a total of six in service with the FAC. In 1987, the FAC began converting some Douglas C-47s into gunships, and, in 1993, the conversion of three into a BT-67 was contracted with Basler Turbo Conversion. The aircraft also received a Star SAFIRE FLIR under the nose, Omega navigation system, an internal communication system and new communications equipment to have contact with the ground troops and the command post. The original M3 12.7mm machine guns were later replaced by GAU-19 miniguns, but sometimes the Nexter M-621 20mm gun is also used. They also carry LUU-2D/B flares for night illumination and LUU-19B/B for use with night vision goggles (NVG).

The unit also has the last IAI 201 Arava within the force, from three received in 1980, which is used for light transport and special forces insertion, and two Cessna 208B Grand Caravans for liaison and one for medical evacuation. Since 1998, the FAC have received 23 Cessna Caravans (five 208 Caravans and 18 208B Grand Caravans), of which 17 currently survive – two 208s and 15 208Bs. These are distributed over the bases, with three being equipped for medical evacuation, one for surveillance (equipped with an FLIR), and the others for transport.

Escuadrón Defensa Aérea 114 (Air Defence Squadron 114)

This is the smallest unit of CACOM-1, but also the one with the rarest of the FAC's aircraft – the only Airbus ECN-235-100M 'Phobos'. This aircraft was originally one of two (the other crashed on 31 July 2015), which by 2014 had been extensively modified, with support from Elbit, to have the Air Keeper solution: electronic support measures/electronic intelligence (ESM/ELINT), electronic counter measures (ECM), communications intelligence (COMINT), communications jamming (COMJAM) and Command and Control (C2) – all of which were operated and managed by a single entity on board a platform. The squadron also has an unmodified CN-235-100M for transport, which is currently out of service. The three CN-235s were received from 1997 onwards.

Above left: The AC-47 was nicknamed 'Fantasma' because it usually operates at night, and the low noise of its engines makes it hard to hear until it is almost overhead.

Above right: A Fantasma, armed with a single GAU-19 minigun. Depending on the mission and needs, these aircraft fly with one or two miniguns, and sometimes they use a 20mm gun.

Below: When more manoeuvrability is needed, the Kfirs use the C12 noses, without the radar. Here, a Kfir C12 stands together with the type that could be its replacement – the F-16 Fighting Falcon.

Above: A Kfir, equipped with the nose of the C12 version, is taxiing after a flight in clean configuration.

Left: In 2021, the FAC received the first Beechcraft T-6 Texans to replace the old Cessna T-37 Tweets, which were retired. Here, one flies together with its competitor, the EMB-314 Super Tucano, in July 2021. (Photo courtesy of the FAC)

Below: The IAI 201 Arava proved a great platform for the short and rough runways of Colombia, but its slow speed is a problem when flying over long distances.

Escuadrón de Combate 116

Until mid-2021, this unit operated the last T-37B Tweets, which are now being replaced by a first batch of three T-6C Texan IIs, with the first two arriving in Colombia in July 2021 and the third being expected in April 2022. These will be followed by a second batch of three, and then a follow-on purchase of six is planned. The intention of the FAC is to reach a total of 24 Texan IIs.

This unit provides training for all future combat pilots of the FAC, with the pupils coming from the CACOM-2 after flying the Tucano. After completing the fighter pilot course at Escuadrón 116, students move onto combat units to fly the Super Tucano, A-37 or Kfir. The FAC received its first 15 T-37Cs in 1968, followed by another three in 1977 and 12 T-37Bs in 1992.

CACOM-2

This command is located at BA Luis F. Gómez Niño at Apiay. It was originally created at San José del Guaviare in 1934 and moved to the current location in 1947. In 1971, the base became CACOM-2, with the main mission of training the fighter pilots, operating the Lockheed T-33s and the Cessna T-37s. Later, the unit operated a mix of counterinsurgency aircraft, including the IA-58 Pucará and the OV-10 Bronco, while Tucanos replaced the retired T-33s. The command has the Grupo de Combate 21, with five squadrons.

Above: A Schweizer SA 2-37B 'Vampiro' of Escuadrón de Combate Táctico 213 (Tactical Combat Squadron 213) is taking off. The aircraft is very capable of finding and tracking guerrilla forces, thanks to its low noise and slow speed.

Right: A Super Tucano taxies at Apiay for a training mission. The type replaced the OV-10 Bronco but proved to be less capable in terms of endurance, armour and weapons capacity.

Escuadrón de Combate 211 'Grifos' ('Griffin')

With the aim of replacing the OV-10 Bronco, which had been in service since 1991, the FAC received its first Super Tucanos on 14 December 2006 from a total of 25 purchased for service within the Escuadrones de Combate 211, 312 and 611. A second batch of 25, to replace the A-37Bs, was planned but later cancelled. Currently, 23 Super Tucanos are still operational with the force, distributed between the three squadrons and constantly performing attack missions against guerrilla groups in Colombian territory. For some years, they used Lizard kits for laser-guided bombs, which were replaced by Paveway IIs, but they mostly use the Mk.81 and 82 bombs without the guiding kits in most missions, as well as 70mm rockets. The Super Tucano was a big step forward in technology for all-weather and night operations, including a weapons delivery system, head-up display (HUD), multifunction displays, a datalink and chaff/flare dispensers. They use an FLIR Star SAFIRE, which is installed in place of the ventral pylon when needed. Besides bombing operations against guerrilla groups, the Super Tucanos are widely used to intercept illegal flights and, in some cases, shooting down unwanted aircraft, as well as destroying some on the ground.

Escuadrón de Combate 212 'Tucanos' ('Toucans')

In 1991, the FAC acquired 14 Tucanos, and these were delivered from December of 1992, replacing the T-33s. While these aircraft always operated from Apiay, they were initially dependant to the Escuela Militar de Aviación, until, in 1994, they formed the Escuadrilla Avanzada 212 (Advanced Squadron 212) of CACOM-2. By then, they served as advanced training aircraft for the pupils that had left the Escuela Militar de Aviación and also started to take part on combat missions against guerrilla groups, using 12.7mm machine gun pods and 70mm rockets. Before the arrival of the Super Tucanos, they were used against illegal flights with machine guns. For some time, a few aircraft were detached to the Escuadrón de Combate 611 of CACOM-6 at BA Tres Esquinas (Caquetá), but now all are based at Apiay.

Over the years, no Tucanos have been lost to accidents or to enemy action, and they were modernised between 2013 and 2020. Improvements included: new wings, new engine mounting, hydraulics, electrical, fuel and braking systems, new landing gear, reinforcement of the fuselage horizontal stabiliser joint, change of the weapon systems harnesses, new digital avionics with two multifunction displays in each pilot position, new DAUs (Data Acquisition Unit) and an ELT (Emergency Locator Transmitter). The delivery of the first upgraded example occurred in July 2014, but problems in the provision of the screens, which were discontinued during the process, delays in the deliveries of the new structural components, and problems with the cockpit canopies, delayed the programme, which ended only on 30 June 2020. This modernisation was intended to extend the operational life of the Tucanos for at least 20 years.

Below left: The Escuadrón de Combate 212 (Combat Squadron 212) 'Tucanos' still has its fleet of 14 Embraer Tucanos, purchased in 1991 and used for advanced training and light attack.

Below right: A Fantasma of Escuadrón de Combate Táctico 213 takes off from Apiay.

A CASA C-212-300 at Apiay. The type is distributed over many units for use on light transport duties.

One of the three Cessna 208B Grand Caravans at Apiay is used for medical evacuation.

Escuadrón de Combate Táctico 213

This squadron has a mixture of aircraft, including two AH-60L Arpías, which are the armed variants of Black Hawks. The Arpía project was born in 1995 as the XM95 Project, installing the ESSS (External Stores Support System) to accommodate two M-3P .50 machine guns and two LAU-61 rocket launchers. In the following years, all 14 received a weather radar to improve the navigation capabilities, and the Arpías changed the M-3P .50s for GAU-19s, plus two M-3Ps for self-protection on the doors and chaff/flare dispensers; aircraft with these modifications became Arpía IIs.

In the early 2000s, the Arpía III was developed, adding an Elbit Systems Toplite II, which has FLIR, Daytime TV, laser designator and rangefinder capabilities for search and weapon designation, EADS AN/AAR-60 MILDS (Missile Launch Detection System), Elbit Systems MIDASH (Modular Integrated Display and Sight Helmet), and GAU-19s replaced the M-3P as door guns (GAU-17 7.62mm guns are used on occasion). Also, new weapons were tested, like the Delilah HL long-range missile developed by Israel Military Industries and the Nexter NC-621 20mm gun, but only a few of the latter were added to the weapons inventory of the Arpías. At least two of the Arpía IIIs received the Elbit PAWS IR (passive airborne warning system) instead of the MILDS. In 2013, the Arpía IV was developed, and the entire fleet of 13 Arpías is being modernised to this standard with the support of Rafael, with Toplite III, Global Link data computer, ANVIS/HUD-24 helmet-mounted sight in place of the MIDASH, a new multifunction display for the use of the Toplite III, and Spike LR, Spike ER and NLOS anti-tank missiles added to its inventory of weapons.

The squadron also has a Cessna 208 Grand Caravan for intelligence, one for medical evacuation and one for transport. Plus, one of the four CASA C-212-300s received for transport since 1989, one AC-47T Fantasma and six Schweizer SA 2-37B 'Vampiro' intelligence aircraft. The latter were initially used by the US Department of State Air Wing (DoSAW) for surveillance, and between 1998 and 2007 were transferred to the FAC. They are equipped with a Star SAFIRE FLIR and other equipment for communications and surveillance. They also boast a high level of endurance, which is fundamental for keeping an eye on guerrilla forces or drug traffickers.

Escuadrón Defensa Aérea 214 'Fénix' ('Phoenix')
This unit has one of the FAC's four Cessna SR-560 Horus airborne early warning and control (AEW&C) aircraft, which is the Cessna OT-47B Tracker (based on the Citation V). Between 2002 and 2004, five of these aircraft were provided to the FAC by the US DoSAW, although one was later sold to a civil operator in the US. The Tracker received an APG-66 radar, extra communications equipment and other systems to provide a limited early warning and command and control capacity to follow illegal aircraft flying at low altitude. The force has a plan to buy new AEW&C aircraft, although this has been delayed by lack of funds. The aim was to buy a platform with an AESA radar, at least eight hours of endurance, air refuelling capacity, 360 degrees of coverage, ECM, satellite communications and datalink.

In 2021, the Tucanos were completely modernised, aiming to extend their service life until the 2030s.

The SA 2-37B Vampiro is one of the strangest aircraft in service with the FAC, but it is used widely for intelligence and surveillance missions.

The Hermes 450 and 900 UAVs (unmanned aerial vehicles) are operated in a secretive way, and it is hard to see them in the open.

Escuadrón de Combate 217 'Quimera' ('Chimera')

This is the main UAV (unmanned aerial vehicle) unit of the FAC, and it operates a fleet of Elbit Systems Hermes 450s and Hermes 900s. The purchase of the systems was made in 2012 for about US$50m, and in 2014, six Hermes 450s and eight 900s arrived, being used on border surveillance and missions against the guerrillas and drug traffickers.

CACOM-3

The origins of this unit are in the Escuadrón de Apoyo Táctico Aero Naval (Air-Naval Tactical Support Squadron), created in 1968 in the city of Cartagena but dependant on the BA de Apiay. Shortly after its creation, it was replaced by the deployment of units of CACOM-1 to the nearby city of Barranquilla, and, in 1977, the Grupo Aéreo del Norte (Northern Air Group) was created at the airport of the city, equipped with light aircraft, Douglas C-47s and helicopters. On 6 August 1978, Grupo Aéreo del Norte became CACOM-3 'MG. Alberto Pauwels Rodríguez', based at BA Barranquilla, and it was equipped with T-33s and, from 1987, A-37Bs. Currently, the unit has the Grupo de Combate 31, formed by four squadrons.

Escuadrón de Combate 311 'Dragones' ('Dragons')

This unit holds the entire fleet of OA/A-37B Dragonflies, which started to arrive to Colombia in 1977, with a first batch of six aircraft. In 1980, they were followed by another 20 and, in 1989, by another eight, most of them being OA-37B standard. In 2001, to reinforce the fleet and replace losses, six A-37Bs that had belonged to the Chilean Air Force were received, though only four were put into service, and, in 2016, two OA-37Bs arrived that had previously belonged to the Dominican Air Force. Until 1987, the Dragonflies served with the CACOM-1, but then they were transferred to Barranquilla to serve on counterinsurgency and anti-narcotic operations over the northern part of the country and the Caribbean.

These aircraft use free fall bombs and rockets and have received some minor modernisations to their avionics, which include a display for the navigation equipment on the co-pilot's panel. Currently,

Above left: A Cessna A-37B of Escuadrón de Combate 311 'Dragones' ('Dragons') taking off from Base Aérea Naval (Naval Air Base) in Brazil during the *Cruzex 2013* exercise, in which the Colombians made their first participation with aircraft.

Above right: The FAC's A-37Bs have received some minor modifications on their avionics over the years, as well as receiving chaff/flare dispensers.

The Escuadrón de Combate 312 'Drako' ('Dragon') operates part of the fleet of Super Tucanos, to support the A-37s in the north of the country.

because of their age and problems getting the cartridges for the ejection seats, only six are on strength (four from the batch from Chile and two from the Dominican Republic), but, in December 2020, their operation was restricted due to further problems with the ejection seats. While options to replace the seats, as Uruguay did in 2017, are being studied, the FAC was, as of 2010, analysing different options to replace the fleet with LIFT type. It studied the BAE Systems Hawk, the Aero L-159NG and the Leonardo M346, but recently the interest went to the KAI FA-50, as it is an aircraft that has some commonalities with the F-16 that the force wants to buy to replace the Kfirs. The idea is not only to have an aircraft that could perform attack missions against guerrilla groups, but also one that could provide air defence to counter the threat posed by Venezuela.

In recent years, Il-96-400VPU intelligence aircraft and Tupolev Tu-160 Blackjack bombers violated Colombian airspace over the northern coast of the country, while flying between Venezuela and Nicaragua. To intercept them, IAI Kfirs were sent from Palanquero, but that base is about 600km from the Caribbean coast, and sometimes they do not arrive in time to make the interception. This has led to the force wanting supersonic aircraft at Barranquilla, so that it can carry out interceptions from there.

Escuadrón de Combate 312 'Drako' ('Dragon')
To support the operations of the A-37Bs, in 2007, part of the fleet of the Super Tucanos purchased from Embraer was delivered directly from the factory to CACOM-3, to create this new squadron. The aircraft are sometimes exchanged with the other four squadrons when they go for maintenance, so the fleet numbers of each change regularly. Currently, the main mission of these aircraft is to fight against the guerrillas in the swamps and hills of the Darien, near the border with Panama, and in the area of Santa Marta, to the east of the base. Additionally, the Super Tucanos are also used to intercept slow-moving aircraft suspected of carrying drugs.

Escuadrón de Combate Táctico 313
This unit is in charge of providing logistic and search and rescue (SAR) support to the command, equipped with a 208B Grand Caravan, two Bell 212s, one Beechcraft C90GTx and two Embraer EMB-110 Bandeirantes, all of which were received in December 1992.

Above and below: The FAC still uses a single Embraer EMB-110 Bandeirante at Barranquilla, for liaison and light transport.

Escuadrón Defensa Aérea 314

This unit is equipped with one or two SR-560 Horuses, according to the availability of aircraft and the needs of the squadrons. These aircraft primarily perform AEW&C missions; however, the aircraft and its radar have also proved very useful in the detection of the so-called 'go-fast' speed boats, which carry drugs from Colombia to the north in the Caribbean. The aircraft perform patrols over the sea and, once they detect a suspected boat, they guide the ARC patrol vessels to intercept them with their helicopters and fast boats.

CACOM-4

Appropriately for the main helicopter command of the FAC, this unit was born as Base de Helicópteros in 1954, when the land to build the base in the town of Melgar, Tolima, about 70km to the southwest of Bogotá, was purchased by the FAC, and the first Hiller OH-12 and OH-23 (originally purchased in 1952 by the Ministerio de Obras Públicas [Public Works Ministry]) and five Bell OH-13Ds, received from 1954, were based there. Since then, the unit has had the task of training the new helicopter pilots of the force. The first medium helicopters were six Kaman HH-43 Huskies, the only of their type in Latin America, delivered in 1961, followed in 1963 by ten UH-1Bs, the first Hueys to operate in a Latin American force. In 1963, the base's name was changed to 'TC. Luis Francisco Pinto Parra', and with the increase on the internal problems with guerrillas, the unit was divided, with the creation of the Grupo Aero-Táctico (GRAT) at Neiva, but both were reunited again in 1964 with the creation of the Comando Aéreo de Apoyo Táctico (Tactical Support Air Command). In 1996, the unit became the Comando Aéreo de Apoyo Táctico No 1 (CAATA-1), and on December 2001, it changed its name again, becoming CACOM-4. In 2002, the part of the command dedicated to training new helicopter pilots was transformed into the Escuela de Pilotos de Helicópteros de la Fuerza Pública (Public Force Helicopter Pilot School), officially training the pilots of the other forces and the PNC.

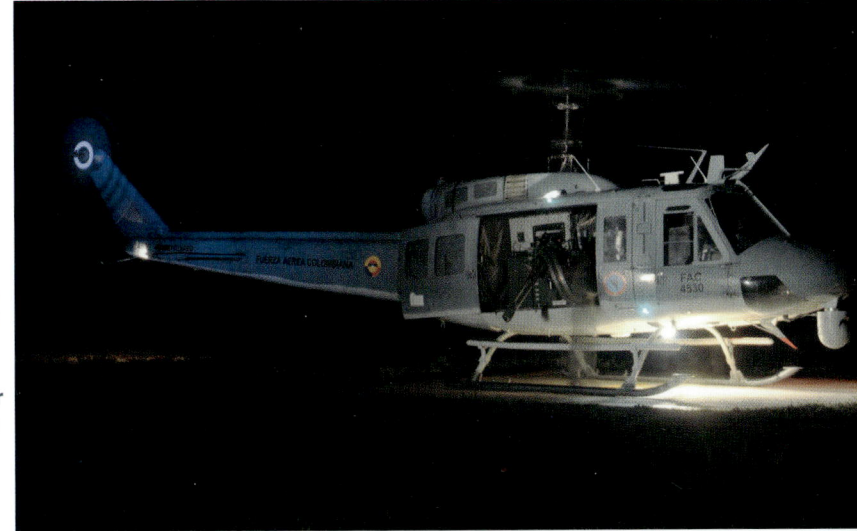

Right: One of the Huey IIs of Escuadrón de Asalto Aéreo 412 (Air Assault Squadron 412) from the batch received in 2008. It is equipped with a Rafael Toplite nose turret and armed with a GAU-19 machine gun on one door and a GAU-17 on the other.

Below: A Huey II flying low over a river, with the GAU-19 visible on the door. They are among the most powerful weapons installed on the Colombian Hueys.

Currently, CACOM-4 has the Grupo de Combate 41, with Escuadrones 411, 412 and 413, and the Escuela de Helicópteros de las Fuerzas Armadas 'Coronel Carlos Alberto Gutierrez' (previously the Escuela de Helicópteros de la Fuerza Pública), complete with a training squadron. Besides the combat missions of Grupo de Combate 41, which comprise transport, assault, combat search and rescue (C-SAR) and escort, the unit also perform humanitarian relief operations following natural disasters and firefighting with Bambi Buckets.

Escuadrón de Combate 411 'Rapaz' ('Rapacious')

This unit, as the parent unit of the Bell 212, is equipped with the last three Bell 212s and four CH-135s, which in Colombia were both baptised as 'Rapaz' and could be equipped with LAU-68A 70mm rocket launchers with seven tubes, plus M-60 or GAU-17 7.62mm machine guns, GAU-19 12.7mm machine guns or M-19 grenade launchers on the doors. Although the Rapazs have previously performed attack missions, now, due to the availability of the Arpías, they are mainly used for C-SAR operations and, in this case, they are usually armed with door-mounted M-60 machine guns and equipped with a rescue hoist.

The first Bell 212 of the force was bought in 1971 for VIP flights, serving with the Escuadrón Presidencial (Presidential Squadron) at CATAM, but in 1984 it was joined by another six. In 1995, the FAC received nine former Canadian Air Force CH-135s and currently most of the survivors serve with this squadron, together with the remaining Bell 212s.

Currently, there is a plan to replace the whole fleet of Bell 212s, UH-1Hs and Huey IIs used across the Fuerzas Militares de Colombia with a single model, with an aim to buy around 100 of the chosen type. The Leonardo AW139 is considered the most interesting one, as it excels in terms of size, performance and operational costs, but a decision is yet to be taken, and the budget has still not been assigned to this.

Escuadrón de Asalto Aéreo 412 (Air Assault Squadron 412)

The unit has the entire CACOM-4 Huey fleet, with ten UH-1Hs and 14 Huey IIs. In 1968, the unit received the first 20 UH-1Hs, followed in 1972 by 20 Bell 205As, which were armed to fulfil attack

A Bell 212 'Rapaz' in flight, armed with M60 machine guns on the doors.

Above: The GAU-19 is clearly seen on the door of this Huey II 'Búho' ('Owl'). They mainly perform their operations during the night, which has led to their nickname.

Right: One of the Bell UH-1Hs still operational with the FAC.

Below: A flight line at Melgar. The base has more than 80 helicopters on strength and flies more than 20,000 hours per year.

missions. In 1982, another 20 UH-1Hs were received, and between 1988 and 1992, the unit also operated the first Black Hawks of the force, before they were transferred to CACOM-5 in Rionegro, Antioquia, about 50km from the city of Medellín. On 7 November 2001, another 17 UH-1Hs, and one UH-1V, arrived as part of *Plan Colombia*, while the last addition was in 2008 and comprised 12 Huey IIs that have the longer nose and a turret for optical sensors. In total, the FAC has received 57 UH-1Hs, 12 Huey IIs and one UH-1V, of which about 40 survive today, and these have been distributed to CACOM-6 and the Grupo Aéreo del Oriente (Eastern Air Group). Of the 14 Huey IIs at Escuadrón 412, two were among the last three modernised between 2013 and 2016, and they are used for air assault and training missions together with the UH-1Hs. The other 12 Huey IIs that did not undergo modernisation are from a batch received in 2008; these feature a Rafael Toplite nose turret with optical sensors, chaff/flare dispensers on the sides of the rear fuselage and usually are armed with a GAU-17 of 7.62mm on one side and a GAU-19 of 12.7mm on the other side on escort missions, or with just two M-60 machine guns for transport and C-SAR operations. As they mostly operate at night, they are called 'Búhos' ('Owls'), and their mission is to escort other helicopters on assault and C-SAR missions during day and night operations.

Above: A GAU-17 is seen on the left door of this Huey II. Usually, these helicopters fly with a GAU-17 on one side and a GAU-19 on the other.

Left: One of the Bell 206Bs used for training at CACOM-4.

Usually, operations against the guerrillas are done at night, as the helicopter crews have the advantage of using NVG. At Melgar, the Huey IIs are mainly used to escort the Bell 212s on C-SAR missions, completing an initial reconnaissance survey of the area where the rescue will take place and then flying over the rescue helicopter. This squadron also has the only fixed-wing assets of CACOM-4, a single Grand Caravan, which is used for liaison and light transport. This aircraft replaced a IAI 201 Arava, which had itself replaced older models, like the Douglas C-47, Beechcraft C-45 and DHC-2 Beaver. The unit also has a single Bell OH-13H that is kept flying for use on exhibitions.

Most of the pilots that finish the helicopter school go to this squadron as their first operational unit.

Escuadrón de Ataque 413 Escorpion (Attack Squadron 413 'Scorpion')
The last squadron of the group is the 413, which only operates the last Hughes 369HS and the MD 500E of the FAC. These helicopters are used for liaison, although in the past they have also been used on scout missions, with the MD 500E being armed with seven-tube rocket launchers and GAU-19 miniguns.

Above left: The last Hughes 369HS with the FAC is one of the oldest of its type still flying in the world.

Above right: The Escuadrón de Asalto Aéreo 412 still flies a Bell OH-13H for exhibitions.

Below: A Rapaz being escorted by a Huey II over the mountains around Melgar.

A Rapaz approaches at low height to a spot between the trees to perform a rescue.

Escuela de Helicópteros de las Fuerzas Armadas 'Coronel Carlos Alberto Gutierrez'

The school is equipped with 11 of the 12 Bell 206B (one was lost) received in 2007 to train the pilots of the FAC, and, until 2016, it had approximately 25 OH-58As (from a batch of 30 delivered by the US government) for use on training of other forces' pilots, including foreigners, mainly from the air forces of Mexico, Honduras, Guatemala and Peru.

On March 2016, the force received ten Bell TH-67 Creeks, which were delivered by the US government to start replacing the old OH-58s, followed by a second batch of 20 in September and another of 30 between 2017 and 2018. Until August 2019, the helicopters belonged to the US government and were loaned to Colombia, but then they were officially transferred to the FAC. Following this, the Regional Helicopter Training Center (RHTC) was opened at Melgar, having originally been held by the US Army at Fort Rucker, Alabama.

The intensity of the training at the school is such that the OH-58A reached more than 25,000 hours of flight in six years, while the Bell 206B reached 30,000 in its first ten years.

As the school performs a high number of flights per year, a plan was developed to move it to Flandes, near the town of Girardot, about 30km from Melgar. However, in the end, only a field was bought close to Flandes, and three small runways were built, inaugurated in 2014, which are used during the day to desaturate the operations in Melgar.

CACOM-5

CACOM-5 is the other main helicopter unit of the FAC, with its base at José María Córdova airport in Rionegro, Antioquia, roughly 50km from the city of Medellín. This unit was organised as BA Rionegro on December 1990, then as Comando Aéreo de Apoyo Táctico No 2 in 1991, finally becoming

Below left: An AH-60L Arpía III taking off. The rocket launchers and GAU-19 miniguns are visible under the wings.

Below right: An UH-60L Black Hawk of CACOM-5 performs a rescue operation using the hoist inside the cabin.

CACOM-5 in 2001. It has Grupo de Combate 51 as its main unit, with three squadrons. The command is the parent unit of the Black Hawk force, having most of the helicopters of the type and performing major maintenance and modifications to them.

Escuadrón de Combate 511 and 512
Both of these units are equipped AH-60L Arpía attack helicopters, all of which are progressively being taken to Arpía IV standard, as described above. They perform attack and escort missions in support of the FAC units but also with the EJC. Furthermore, in addition to attack missions, the Arpías of the 511 have the Equipo Acrobático Arpía (Harpias Aerobatic Team), which perform aerobatic exhibitions, especially for the F-Air airshow organised every two years at Rionegro airport.

Escuadrón de Operaciones Especiales 513 (Special Operations Squadron 513)
This squadron has a 208B Grand Caravan for liaison and the rest of the Black Hawk fleet of CACOM-5. The FAC is the pioneer in the operation of the Black Hawk in Latin America, and it has one of the highest levels of combat experience with the model in the world, using them in combat since the first five UH-60As arrived at BA Tres Esquinas in Palanquero on 14 July 1988. A second batch of five followed in February 1989, and all were pressed into service at (what was) CAATA-1 at BA Melgar.

In 1992, the aircraft were transferred to CAATA-2 at Rionegro, and, at the same time, operations with NVGs began. Two years later, four UH-60Ls arrived, and in 1996, six Bambi Bucket 5566s of 2,500 litres were purchased from SEI Industries in Canada, starting firefighting operations the same year, and which have been continued since. In 2000, seven extra UH-60Ls arrived, followed by another five in 2001. All of the remaining UH-60As were taken to UH-60L standard, and 13 were modified to AH-60L Arpía standard. Nine UH-60L serve with the squadron for transport duties, while another was modified with VIP interiors and serves as the presidential helicopter at CATAM.

An Arpía IV firing a Rafael Spike missile during tests. This version was developed to face Venezuelan armoured units, as a result of growing tensions between the two countries, as well as Colombia's lack of tanks. (Photo courtesy of Rafael)

Above: Two Arpía IIIs approaching. These helicopters are lacking the FLIR turrets and weapons.

Left: A fully armed Arpía III ready for a mission. At CACOM-5, as well as on the other bases with Arpías, at least one or two are permanently on alert.

Below: An Arpía III flies over the hills close to Rionegro Airport. The type proved extremely efficient against guerrilla groups, with a huge firepower and high survivability.

Above: An Arpía III performing a rescue practice using the hoist inside the cabin.

Right: Colombia is one of the biggest operators of the Black Hawk in the world and has some of the most extensive combat experience with it, which has led to the development of the Arpía variant.

Below: An Arpía III firing all of its flares. While the guerrillas never managed to operate MANPADS (Man-portable air-defence systems) successfully, they made some attempts, so all the Colombian military aircraft always fly with flares.

In 2005, the US government donated two MH-60Ls specially equipped for C-SAR missions, which arrived in 2008 and were baptised 'Ángel'. They featured an FLIR turret, radar, rescue hoist, chaff/flare dispensers and first aid equipment on board. They were followed by a third in 2013.

CACOM-6

This unit was originally born as an auxiliary base at the small town of Puerto Boy in the Putumayo region during the conflict with Peru between 1932 and 1933. Officially created in October 1933, in 1939 it was taken to Tres Esquinas, at the junction of the Orteguaza and Caquetá rivers. In 1959, the base was closed due to lack of funds, however, it was reopened in 1966 as Grupo Aéreo del Sur (Southern), with Douglas C-47s, C-45s and Beavers on wheels and floats. In 1990, the base was named 'Capitán Ernesto Esguerra Cubides', and between 1992 and 1993, a radar station was installed as part of a programme to install radars to cover the entire Colombian territory, with other radars in Marandúa, Araracuara, Mecana, Bogotá, Cali, Medellín and Barranquilla, among others.

The base at Tres Esquinas was expanded in the 1990s, with a bigger, better paved runway and other facilities, and, in 2001, it became CACOM-6, with the Grupo de Combate 61, but it still has small facilities, being the smallest of the CACOMs. This unit has only two squadrons, and they borrow aircraft from the parent units of each type, according to their needs.

Escuadrón de Combate 611

This is the third unit within the FAC that operates the Super Tucano for close air support, attack and control of illegal flights over the southeast of the country and over the borders with Ecuador and Peru. It typically operates around three of these aircraft at any given time.

Escuadrón de Combate Táctico 613 'Pelicano' ('Pelican')

This unit operates one of each of the following: a CASA C-212-300, Grand Caravan, SA 2-37B Vampiro, AC-47T Fantasma, Huey II, Bell 212 Rapaz, and also uses some of the Boeing Insitu Scan Eagle UAVs possessed by the FAC.

Above left: The CASA C-212s are used by many units for light transport, including CACOM-6. The aircraft in the picture wears the badge of the Escuela Militar de Aviación (Military Aviation School).

Above right: A Cessna 208 Caravan, one of two that the force still operates, from a total of five received from different sources.

CACOM-7 and the Escuela Militar de Aviación

After the conflict with Peru, it was decided to move the Escuela Militar de Aviación, and, on 21 September 1933, the new facilities were inaugurated at the city of Cali. During World War Two, the school started to receive 20 BT-15 Valiants, 60 PT-17 Kaydets, about 100 AT-6 Texans and ten Fairchild PT-19 Cornells. In 1954, the replacement of these original aircraft began with the Beechcraft T-34A Mentors and, from 1968, with 30 Cessna T-41D Mescaleros. In 1983, the school also received its first gliders, with two ICA IS-28B2 Larks. In 2010, the Mentors were replaced by 25 locally built CIAC T-90 Calimas, which is a license-built version of the Lancair Legacy FG. Around 2013, they were upgraded to T-90C standard, with better avionics and reinforced landing gear, as the use during training was tougher than the original design accounted for.

The Escuela Militar de Aviación receives all those who want to become officers within the FAC, normally between 100 to 150 each year, and instruction lasts for four years. In the third year, they start to fly the T-41D or the T-90C, and, in their fourth year, after 48 to 60 hours of primary training, the pilots progress to basic training with the Texans of CACOM-1, the Tucanos of CACOM-2 or Bell 206 helicopters of CACOM-4, flying there for about 100 hours. After this, they graduate as subtenientes (second lieutenants).

To increase the operational capacity of the force in the western part of the country, in February 2013, CACOM-7 was created within the facilities of the school, receiving aircraft borrowed from other units. The school has the Grupo de Educación Aeroáutica (Aeronautical Education Group), which comprises two squadrons, the 711 and 712, while CACOM-7 has the Grupo de Combate 71 with the Escuadrón 713.

Escuadrón Preparatorio 711 (Preparatory Squadron 711)

This squadron is equipped with two Stemme S10-VT motor gliders, which, in 2016, replaced the two IS-28B2s and the two L-23 Super-Blaniks, which retired in 2018, and still has one SZD-54-2 Perkoz.

A CIAC T-90C Calima, which replaced the Beechcraft T-34 Mentor for training at the Grupo de Educación Aeroáutica (Aeronautical Education Group) of the Escuela Militar de Aviación.

The Escuela Militar de Aviación still has a single Mentor for use on demonstrations.

The Cessna T-41Ds are being replaced by Cessna 172s for basic training.

For primary flying, the unit has nine Cessna T-41Ds from a batch of 30 received in 1968, which are now being replaced by an undisclosed number of Cessna 172Ss, of which the first four arrived in July 2021.

Escuadrón Básico 712 (Basic Squardon 712)
This squadron has the T-90C Calima, of which a total of 25 were originally built by CIAC between 2010 and 2015, though one was lost in an accident in 2017. It also operates a single PT-17 Kaydet and a T-34A Mentor as part of its historical flight division.

Escuadrón de Combate Táctico 713
Like other similar units from other CACOMs, this squadron operates aircraft usually sent by the parent squadrons of each type. In this case, this squadron operates one A-29B Super Tucano, one AC-47T, one Arpía IV, one Bell 212 Rapaz, one Grand Caravan and one CASA C-212-300.

CATAM

This is the main transport unit of the FAC, having four main groups in charge of different missions. The organisation has its origins on the Escuadrón 101 de Transporte, created in 1944 at BA Madrid with a mixture of German and US aircraft. Later, in 1954, the Escuadrón de Enlace (Liaison Squadron) was created at the old airport of Techo in Bogotá, for VIP flights under the orders of the Republic's president. This unit was merged with Escuadrón 101 in 1955, creating the Grupo de Transporte Aéreo Militar (Military Air Transport Group), and in 1959 it became Base de Transportes. Shortly after, the unit moved to the new airport of El Dorado, and on 1 January 1966, it became the Comando Aéreo de Transporte Militar. Currently, the unit is organised into three Air Groups and the Servicio de Aeronavegación a Territorios Nacionales (SATENA, Air Navigation Service to National Territories), an airline run by the force to reach isolated places of the country.

The KC-767 is ready for air refuelling. The aircraft is widely used, and the FAC has an interest in acquiring another one.

The KC-767 firing its flares for the first time in Colombia, in October 2018, over the mountains to the north of Base Aérea (BA) de Germán Olano in Palanquero.

Grupo de Transporte Aéreo 81

This unit holds the tactical and strategic transport aircraft of the FAC, as well as some light transport aircraft. It has Escuadrón de Transporte 811 as its sole squadron. The unit has one KC-767, baptised 'Jupiter', as its biggest asset. The plane was bought in Israel in 2009, to replace a KC-707, and it was converted for air refuelling and cargo, receiving also a radar warning receiver (RWR), chaff/flare launchers, and other systems. While there was interest in buying a second 767, this was not possible due to lack of funds. There is also a Boeing 727-2B7 and 727-2X3F, together with a Boeing 737-4S3SF and 737-46BF, all with cargo doors, providing jet transport capacity.

The tactical transport force includes the fleet of C-130 Hercules aircraft. In 1968, three C-130Bs were acquired, of which the first, with the serial number FAC1001, is still operational. In 1983, two C-130Hs were received, and these are also still operational. However, eight C-130Bs, donated by the US since 1987, have been retired. To expand the fleet, one C-130H arrived in 2004, and in 2020 three more were negotiated from US surplus, with one arriving in 2020 and two in August 2021.

As the FAC was no longer using the Douglas C-47 for transport, and after the retirement of the C-130Bs, new lighter aircraft were needed. In 2007, four Airbus C295Ms were ordered, followed by another two in 2012, as it was considered that there were many missions for which the Hercules was too big, and the operational costs for the C295M were 40 per cent less. For light transport, the unit has three Beechcraft C90GTxs, three Beechcraft 350s for medical evacuation and use of all those Grand Caravans that are not deployed to other units.

Grupo de Vuelos Especiales 82 (Special Flights Group 82)

This is one of the special units of the force, and it is in charge of the VIP flights for the national authorities. It has one squadron, which is the Escuadrón de Transporte Especial 821 (Special Transport Squadron 821). The main presidential aircraft is a single Boeing Business Jet (737-74V), and there is also one Fokker F 28 Mk1000, one Mk3000C, two Embraer ERJ135BJs, two Bombardier Learjet 60s and one Cessna 525A CitationJet. The unit also has rotary-wing assets, with one Leonardo AW139 received in 2021 to replace a lost Bell 412EP, one UH-60L Black Hawk, modified with VIP interiors, and one Bell 412EP.

The force still uses two Boeing 727s for cargo and passenger transport.

Grupo de Inteligencia Aérea 83 (Air Intelligence Group 83)

Together with Escuadrón 114 of CACOM-1, this is the electronic intelligence unit of the FAC, having the Escuadrón de Inteligencia Aérea 831 and being equipped with two Beechcraft 350 Super King Airs, one Beechcraft 300 and one Aerocommander 695 modified with ELINT equipment. This unit and its aircraft have proved fundamental in the war against guerrilla groups and drug traffickers, detecting and intercepting their communications.

Servicio de Aeronavegación a Territorios Nacionales (SATENA)

The geography of Colombia makes travel by land very complicated and, in some areas, impossible, especially with the Amazon, but also on the highlands of the Andes. With very poor road infrastructure, air travel has always been very important in the country, and, in 1919, one of the first airlines in the world, SCADTA, was created. However, air transport was only available to a small part of the population, and those living in the most isolated places could not afford it. To reach those places, in 1943, the government created an air service to carry passengers and cargo to isolated places in Colombia, using the aircraft of the Escuadrón 101 de Transporte. On 12 April 1962, SATENA was created, having its own fleet of passenger and cargo aircraft, including the Douglas C-47, C-54, PBY-5 Catalina and Beaver. Over the years, the airline received more C-47s and added a Fokker F 28, Avro HS-748, Pilatus PC-6 Turbo Porter and CASA C-212-100. In more recent years, it has operated Dornier 328s, Douglas DC-9s, Harbin Y-12s, Embraer EMB-120 Brasilias, Embraer ERJ-170s and Boeing 727s, all of which are now retired. Currently, the fleet comprises two ATR 42-512s, two ATR 42-500s, three ATR 42-600s and two ERJ135BJs. Although SATENA operate like an airline, the aircraft have both civil and military serials, and the crews are military.

The Boeing 737-46BF used for passenger and cargo transport by the Comando Aéreo de Transporte (CATAM, Air Transport Command).

Above left: Each jet transport of CATAM has its own name, with this called *Cronos*.

Above right: One of the two Boeing 737-400s of the FAC prepares for take-off from Tolemaida Aeródromo Military (Army airfield).

The KC-767 is escorted by four Kfirs, including a C12, a C10 and two TC12s.

Above left: One of the Lockheed C-130H Herculeses of the Escuadrón de Transporte 811.

Above right: FAC1008 is a C-130B and operated until around 2020, when it was grounded at CATAM. It is going to be preserved at the base.

FAC1014 is one of the C-130Bs operated by the force and now grounded, being replaced in 2020 by three C-130Hs delivered by the US Air Force.

The Airbus C-295s reinforced the fleet of C-130s and were a replacement for the Douglas C-47s.

Above and left: In 2007, the FAC purchased four Airbus C-295s for use by CATAM in medium transport missions.

The Escuadrón de Transporte Especial 821 (Special Transport Squadron 821) of CATAM uses an Embraer ERJ135BJ Legacy for VIP transport.

Above: For use as a presidential aircraft, the Escuadrón de Transporte Especial 821 has this Boeing Business Jet.

Right: The Legacy is sometimes used for presidential flights within Colombia, but it is mostly for other members of the government.

Above: The Escuadrón de Transporte Especial 821 still uses two Fokker F 28s, being one of the last two Latin American operators of the type, the other being the Argentine Air Force.

Left: The Escuadrón de Transporte Especial 821 operated two Bell 412EPs for VIP flights, but the example in the picture was lost in an accident on 25 October 2019. It was replaced by a Leonardo AW139.

Below left: One of the Black Hawks of CACOM-5 was modified for VIP transport and transferred to Escuadrón de Transporte Especial 821 with a new serial. On 25 June 2021, the helicopter received gunfire from guerrilla forces while carrying the Colombian president to the city of Cúcuta. It was hit three times but landed safely.

Below right: One of the rarest planes of CATAM is this Beechcraft 350 Super King Air of Escuadrón de Inteligencia Aérea 831 (Air Intelligence Squadron 831), equipped with satellite communications (SATCOM) antenna and electronic intelligence (ELINT) equipment, and wearing a civil paint scheme.

Grupos Aéreos

Grupos Aéreos are smaller in size than a CACOM, with only one flying squadron and, in most cases, a few aircraft are rotated between units according to need. They also receive other units when they are deployed for specific missions, and the Grupos Aéreos provide support.

Grupo Aéreo del Amazonas
At the southeast end of Colombia, at the border with Peru and Brazil, is the town of Leticia, which can only be reached by air or by river, so air operations there are vital. It is close to areas where guerrilla groups operate, and the proximity to the borders means it is an area they must control. Leticia is home to the Grupo Aéreo del Amazonas, with the Escuadrón de Combate 401, and is equipped with a single CASA C-212-300 and a Grand Caravan.

Grupo Aéreo del Oriente
This unit is based at Marandúa, to the east of Colombia, in the grasslands and rivers of the Orinoco basin, close to Venezuela. It was created in the early 1990s, as there was a lot of guerrilla activity in the area. The unit has the Escuadron de Combate Táctico 201, equipped with Bell 212 Rapazs, Huey IIs and Cessna 182Rs, for liaison, and Scan Eagle UAVs.

Grupo Aéreo del Caribe
The island of San Andrés, in the middle of the Caribbean, is a strategically important place for Colombia. On one side, the archipelago of San Andrés and Providencia were in dispute with Nicaragua for many years, and the Fuerzas Militares de Colombia has maintained some permanent military presence in the area since 1979, when Nicaraguan aircraft started to violate the Colombian air space over the area. Between 1979 and 1987, the island only received aircraft from other units, including A-37s and Mirage 5COAs. However, on 3 June 1987, the BA San Andrés was created, having four A-37s on strength until the late 1990s.

Some of the drugs taken from Colombia are sent to the Caribbean on fast boats (called 'go-fast' boats), and San Andrés is close to the routes followed by those boats, being an ideal place for the surveillance aircraft to operate. The Grupo Aéreo del Caribe has the Escuadrón de Combate 101, which has the Escuadrilla de Combate Táctico 1013 and a single Beechcraft C90GTx. However, the unit constantly receives aircraft from other units from the mainland, including Kfirs, Super Tucanos, AC-47Ts and more A-37s.

Grupo Aéreo del Casanare
This is the latest base to be organised, at Yopal, in the department of Casanare, in the centre-east of the country, on an area of grasslands close to the border with Venezuela, where guerrilla groups have operated for many years. It was inaugurated in 2010, and for the next five years it operated the remaining OV-10 Broncos, until the type was retired. Now, the Escuadrón de Combate 301 operates some Super Tucanos, Arpía IVs, Grand Caravans and Scan Eagle UAVs. The quantity of aircraft varies according to the availability and need.

A Beech 350C, which is equipped for medical evacuation, used by the Escuadrón de Inteligencia Aérea 831.

The smaller units of the force mainly use a variety of liaison and small transport aircraft, the Grand Caravan being one of the most common.

CAMAN

CAMAN is located at Base Arsenal Madrid, to the west of Bogotá, and is in charge of all the major maintenance of the aircraft of the force, as well as any modification or modernisation. When the transport units located at Madrid were moved to Techo airport in 1955, the base was named Base Arsenal and was dedicated only to maintenance. Later, it was named Comando Aéreo de Material (Material Air Command), and when all the supply units were moved to El Dorado to become part of CATAM, the base became the Comando Aéreo de Mantenimiento (CAMAN).

Grupo de Transporte Aéreo 91

The group performs light transport and liaison missions, having the Escuadrón de Transporte 911 and one CASA C-212-300.

Corporación de la Industria Aeronáutica Colombiana (CIAC, Colombian Aeronautical Industry Corporation)

CIAC is located within the same facilities as CAMAN, and it is the largest aeronautical facility in Colombia, performing some major modifications to aircraft and, in the past, being in charge of the construction of the T-90 Calima. It is currently developing the Quimbaya light UAV, which, as of 2021, started to be delivered to the FAC for tests.

Fuerza de Tareas Aéreas (FTA, Air Task Force)

This task force was created in 2012 with special forces from the FAC, as well as from other military branches, to act against drug traffickers on the departments of Arauca, Vichada and Guainía. It comprises a company from the Brigada Especial Antinarcóticos (Special Brigade Against Drug Trafficking) and one Compañía Operacional de la Dirección Antinarcóticos (Operational Company of the Antinarcotics Direction) of the police, a Riverine Assault Group of the Marine Force, communications personnel from the EJC and a staff and intelligence centre with personnel from all the forces.

Grupo de Operaciones Especiales Aéreas (GROEA, Air Special Operations Group)

Based at the CATAM base, this unit has the Escuadrón de Comandos Especiales Aéreos (ECOEA), which is the special forces of the FAC, dedicated to C-SAR, hostage rescue, personnel recovery, management of explosives in case of terrorist actions, forward air control, preparation of forward fields for aircraft and helicopter operations, exploration and direct actions against the enemy.

Chapter 2
Aviación Naval Colombiana – Colombian Naval Aviation

When the war with Peru erupted in 1932, the Colombian government realised the need to develop its military aviation and, with very few airfields in the country but many rivers and coasts, the use of hydroplanes and flying boats became relevant. In 1933, the Escuela de Hidroavión (Hydroplane School) was created at Buenaventura, on the coast of the Pacific Ocean, close to the city of Cali, while the SCADTA airline facilities in Cartagena were requisitioned for use by hydroplanes. In 1937, the transport ship *Cúcuta* was modified to carry two Curtiss Cyclone Falcons, which were lowered to the water with a hoist for take-off, and then recovered the same way after they landed. Operations on this ship began on 4 July 1937 and ended in 1942.

When Colombia joined the Allied side in World War Two in 1942, declaring war on the Axis countries, anti-submarine patrols were performed using Dornier Wals and Consolidated P2Y-1C patrol flying boats, but it was still dependant to the Dirección General de Aviación Militar (General Directorate of Military Aviation) of the EJC. The air base at Cartagena was later transferred to the ARC, becoming Base Naval (BN) ARC Bolívar, but the aircraft were still operated by the EJC. However, in January 1944, the ARC gained its first pilot, Teniente de Navío (Ship Lieutenant) Carlos Quijano, who graduated at the Comando de la Aviación Naval Argentina (Argentine Naval Aviation Command). By this time, the ARC planned to develop its own aviation command, and Aviación Naval (Naval Aviation) sections were included in the plans to build a naval base on the Pacific Coast and one at Barranquilla. However, these plans did not come to fruition, and the ARC would have to wait to have its own aviation.

In 1968, the Patrulla Aérea Civil de Bogotá (Bogotá Civil Air Patrol), a civil organisation, offered the Fuerza Naval del Atlántico (Atlantic Naval Force) of the ARC the support of civil aircraft for coastal maritime surveillance operations, using light aircraft from private operators. In 1969, Patrulla Aérea Civil de Bogotá took part in the Operation *Halcón Vista* (*Hawk View*) with the ARC with Piper PA-28 Cherokees, something repeated in the following years, until, in 1972, the Patrulla Aeronaval de Cartagena (Naval Air Patrol of Cartagena) was created; this was also a civil organisation, but it was specifically directed to support the ARC. In 1975, with support of the ARC, a flying school was created for the Patrulla Aeronaval, which trained both civilians and ARC officers as aviators. This organisation purchased some aircraft, including PA-28s, Cessna 172s, Aero Commander RC560s and RC690s and a Hughes 300C helicopter.

When, in the early 1980s, the ARC purchased four Almirante Padilla-class corvettes, equipped with flight deck and hangar, the force recognised it needed to supply and utilise these areas so it could perform embarked operations, and that better organisation was needed. Because of this, in September 1983, the ARC decided to create the Comando de Aviación Naval, with its command at Bogotá and, together with the corvettes, two MBB Bo 105 helicopters were acquired. Furthermore, two Aero Commander 560s and three PA-28s of the Patrulla Aeronaval were transferred to the command, as well as its facilities in Cartagena, where the Escuadrón de Ala Fija (Fixed-Wing Squadron) was created, while the helicopters, forming the Escuadrón de Ala Rotativa (Rotary-Wing Squadron), operated from NB ARC Bolívar, also in Cartagena. On 1 January 1987, the Grupo Aeronaval del Atlántico (Atlantic Naval Air Group) was created.

To have presence over the Pacific Ocean, in 1977 the ARC created the Grupo Aeronaval del Pacífico (GANPA, Pacific) in the town of Juanchaco, near the city of Buenaventura and close to BN Bahía Málaga. In that same year, two AS555SN Fennec helicopters were purchased for embarked operations, as well as five Bell 412s and one Bell 212 to support the activities of the marine forces against guerrilla groups and drug traffickers, both on the Pacific and Caribbean coasts and on the rivers across the country. The fixed-wing aircraft grew with the addition of a Beechcraft 350 Super King Air, a Grumman Gulfstream I and a locally built Gavilán G-358, the last for operations in the east of the country in support of the marine forces on the rivers, based at Puerto Carreño.

For the operations on the rivers, the Grupo Aeronaval de Transporte y Apoyo Fluvial (GATAF, Naval Air Transport and River Support Group) was created in 2001, based at El Dorado airport in Bogotá, which deployed a Bell 412 to the south of the country, later joined by the Bell 212 and a Cessna 206, in support of ground operations by the marine force. The Cessna 206 replaced the Gavilán at Puerto Carreño. With the increase of activities of drug traffickers using the so-called 'go-fast' boats, in 2003 the first two CN-235s were purchased from EADS CASA. These were -100 aircraft, transformed into -200 and modified for maritime patrol, with an RDR-1500 radar and a Wescam 15 optical sensors turret, plus a launcher for illumination flares and two observation bubble windows. By this time, the Grupo Aeronaval del Atlántico was renamed Grupo Aeronaval del Caribe (GANCA).

Since, in 2002, the Comando General de las Fuerzas Militares (General Command of the Military Forces) ordered that all military pilots of the Fuerzas Militares de Colombia would be trained at the Escuela Militar de Aviación in Cali, with the helicopter pilots being trained at the CACOM-4 of the FAC, the Comando de Aviación Naval closed its flight school in 2003.

The fleet continued its modernisation and expansion with the arrival of two Grand Caravans in 2004, joined by a Grand Caravan EX in 2013. Furthermore, in 2003, two PA-28s, one Cessna 152, one Piper PA-31 and two Enstrom F 28 helicopters were exchanged for one Gavilán G-358 and one Cessna 206. One MBB BK117 helicopter was purchased, and in 2004, the Spanish government donated a CASA C-212-100 to the GATAF, which in 2017 was transferred to the EJC. The ARC had started to operate Cessna 206s in the late 1980s, with two impounded aircraft, one of which, ARC-405, is still operational. With examples arriving until 2005, there was a total of seven, but only two survive today.

In 2010, new facilities were inaugurated for GANCA in Barranquilla, where the unit moved its headquarters and most of its aircraft, leaving a small station at Cartagena. Also, in 2008, new facilities were built at Bogotá for the GATAF, while a portable hangar was bought and installed at NB Puerto Leguizamo at the Putumayo River, on the border with Peru.

To expand the operations on riverine areas, in 2009, six Bell CUH-1Ns were transferred by the EJC, for operations on the rivers in support of the marine forces, as part of the GATAF, which in 2021 was renamed Grupo Aeronaval Central (GANCE, Central Aeronaval Group). They were followed by four Bell 412EPs, received between 2013 and 2014, for embarked operations, replacing the retired MBB Bo105 and BK117 and reinforcing the fleet of a single Fennec, three Bell 412s received previously and a single Bell 212. As one Fennec was lost in an accident in 2016, and the other was grounded for a long time, two SA365N2 Dauphins were received in October 2018, one of them purchased by the DIMAR (Dirección General Marítima, General Maritime Directorate) but both for use by the ARC. While the force wanted to receive another two in 2019, this was delayed because the two examples received were plagued with problems, and claims were made to the seller for not providing what was negotiated.

For the future, a main priority for the ARC is the replacement of the UH-1N, for which the government plans to find a common replacement for the entire fleet of Huey IIs, UH-1Hs and UH-1Ns of all the Fuerzas Militares de Colombia and the PNC. The Leonardo AW134 is the preferred choice, but it is also analysing other options, like the Bell 412. The ARC also aims to increase its embarked capabilities, and the intention is to buy more Dauphins, while the ARC has the PES (Plataforma Estratégica de Superficie, Strategic Surface Platform) programme, for the selection of a design for future frigates to replace the Almirante Padilla class. Although the programme is progressing very slowly, the ARC wants to have a more powerful embarked helicopter, with anti-submarine warfare and anti-surface warfare capacity, for which it is analysing options like the Sikorsky Sea Hawk and the Leonardo AW159. The intention is to have five of them, while it also wants to have nine light helicopters like the AS365Ns, two of them belonging to the DIMAR.

The fixed-wing fleet was expanded in 2011 by a 350i Super King Air, to replace the former one, retired in 1999. It was followed in 2015 by a single ATR 42-320 and one Beechcraft C90 in 2018. In the future, the ARC aims to buy at least two new maritime patrol aircraft, like the CASA CN-235-300 MP Persuader, a second ATR 42, a fleet of seven Grand Caravans and three light aircraft, one of them belonging to the DIMAR. In December 2021, a new C90 was added, impounded from drug traffickers, while a Beechcraft 340ER Super King Air was delivered from the factory, as well as a Bell 412EPi.

The ARC has plans to create a new Estación Aeronaval (Naval Air Station) at San Andrés Island, one at Puerto Carreño and one at Puerto Leguízamo, and to have aircraft based there permanently.

Current operations

In 2021, the ARC had a reorganisation, with all the units now depending on three Jefaturas (Commands): Alistamiento (Readiness), Operaciones (Operations) and Personal (Personnel). The Comando de Aviación Naval became Comando de Alistamiento Aeronaval (Naval Air Readiness Command), which is in charge of administering the naval aviation resources (aircraft, personnel, weapons, etc.), while the Comando de Operaciones (Operations Command) has under it the Fuerza Naval del Caribe, del Pacífico, del Sur and del Oriente, and the Grupos Aeronavales that depend operationally on them, with GANCA depending on the Fuerza Naval del Caribe, GANPA on the Fuerza Naval del Pacífico and GANCE operating under the command of the other two, depending on locational need. Each of the three groups has its own area of responsibility and equipment. GANCA and GANPA sometimes exchange aircraft, according to needs and availability.

GANCA

This the biggest of the groups; it is located at Barranquilla, with a station at Cartagena, where the main BN of the ARC is also located. The group is part of the Fuerza Naval del Caribe, on which it depends operationally, but belongs to the Comando de Aviación Naval.

While not a dependant of the group, the Escuela de Aviación Naval (Naval Aviation School) shares the base. This school has no aircraft or crews of its own, instead using those from the squadrons of the group. It is in charge of receiving the helicopter or aircraft pilots from the Joint Schools at Melgar and Cali, respectively, training them on naval aviation operations and giving them the specific training on the aircraft type they will fly.

The group has two squadrons, one of fixed-wing aircraft and one of helicopters. The first is equipped with two CN-235-200s and one Persuader, plus a single Cessna 206 used for liaison and training. The helicopter unit has four Bell 412EPs, one 412EPi, a single CUH-1N, two AS365Ns and one AS555 Fennec (which is based in Cartagena). In 2003, the first two CN-235s were purchased. In 2010, these were joined by the third aircraft, newly built and with the FITS (Fully Integrated Tactical System),

One of the Airbus CN-235-200s purchased in 2003. The aircraft were originally civil CN-235-100s that belonged to Binter airlines, and they were modified for maritime patrol.

equipped with a Telephonics APS-143C V3 synthetic aperture radar and an FLIR Star SAFIRE III. The original two later received the same radar.

Since 2010, all of these aircraft operate from GANCA, but usually one is deployed at GANPA in Juanchaco and one at San Andrés Island in the Caribbean. In tackling drug traffickers, all of these aircraft can perform normal maritime patrol missions of trying to detect any suspicious activity, or they could be scrambled when intelligence information about a possible activity arises. In both cases, the operations are decided and supervised by the Supervisor Operacional (Operational Supervisor), who is in charge of all the operations against drug traffickers in the area of the Caribbean (there is another one for the Pacific).

Traffickers carry drugs over the sea in four different ways. One is using 'go-fast' boats, which are capable of reaching more than 40 knots. The second is using submersibles or mini-submarines, which are hard to detect. Another is on sailboats, which simulate the appearance of normal cruise, and the last one is carrying the drugs on small vessels to cargo ships far from the coast. This last system is

Below left: The single CN-235-300MP Persuader is the most modern maritime patrol aircraft of the force. There is interest in buying more of this type.

Below right: The single CN-235-300MP is equipped with a Telephonics APS-143C V3 synthetic aperture radar and an FLIR Star SAFIRE III.

In 2018, the Armada Nacional de la República de Colombia (ARC, Colombian National Navy) added RAMPAC capacity to the CN-235s, enabling these aircraft to facilitate a rescue team by parachute and a boat to perform rescues far from the coast.

mostly used when the destination is Europe, Asia or Africa. The 'go-fast' boats are easy to detect and to determine as an illegal boat, as they are the only ones sailing at such speeds on the open sea. In most cases, they depart the Colombian coast and head east, towards Venezuelan waters, from where they head north to Puerto Rico, Haiti, Cuba or another island from where the drugs are then taken to the United States. The radar of the CN-235s can detect them at 25 to 30 miles, but the aircraft must keep a distance to be undetected, while co-ordinating with the Centro de Operaciones (Operations Centre), which receives all the information datalink to send surface vessels for interception. Pilots often start filming the boats, as, usually, when they realise they have been detected, the traffickers drop the drugs into the sea. If it is another type of boat, pilots check if it has an AIS (automatic identification system), and if not, they go to investigate and co-ordinate the subsequent interception. The onboard radar of the CN-235 can detect a big ship at up to 200 miles and a small one at 60 miles. They can also detect the snorkel or antennas of the submersibles, and observers can confirm the detection using the bubble

The first CN-235, ARC 801, flying over part of the ARC fleet, which includes a Type 209 submarine (the ARC has two of them in total), one of the two Klasse Lüneburg Type 701A supply ships and two of the Almirante Padilla-class corvettes. (Photo: Colombian Navy)

A CN-235-200 flying over an Almirante Padilla-class corvette. (Photo: Colombian Navy)

Above: A Cessna 206 is assigned to the GANCA for training and liaison missions, and to train paratroopers.

Right: The ARC has a single Airbus AS55SN Fennec, which can be equipped with a Nexter 20mm gun. It is usually embarked on the Almirante Padilla-class corvettes and operates from Cartagena station.

windows. While the pilots do not use NVGs, because they always fly at over 5,000ft, the observers use them to see objects on the sea during the night. The FLIR turret is also used to detect and follow suspicious objects.

Usually, the ships that perform the interceptions are the three offshore patrol vessels (OPVs) of the 20 de Julio class, all Fassmer OPV 80 types, which carry a Bell 412 from GANCA. The helicopters received the same radar as the CN-235s to increase their capacities. While it was planned to equip the Dauphins with the same radar as well as sonobuoys and torpedoes to also have ASW (anti-submarine warfare) capacity, this did not happen. When the OPVs are sent to intercept a suspected vessel, they usually operate with the helicopter and their interceptor boat, which can be deployed from the stern. On board the helicopter, a boarding team of special forces are stationed, ready to descend on rappel or fast rope. In the case of the 'go-fast' boats, the helicopter carries a sniper, armed with a Barrett 12.7mm rifle, to fire against the engines of the boat if the crew refuses to stop. Most of these operations are performed in the night, not only because that is when drug traffickers most often leave the coast, but because the aircraft are less likely to be detected by the boats below.

The unit also provides SAR, and rescuers are deployed both from helicopters and, as RAMPAC (Rescate en Alta Mar mediante el lanzamiento de Paracaidistas y Carga, Sea Rescue by launching Skydivers and Cargo) capacity has been developed, a rescue team can also be deployed from the ramp of the CN-235s, followed by a motorboat to quickly carry the castaway to a rescue vessel. The rescuers jump from 10,000ft, and then the aircraft descends to 500ft to drop the boat as close as possible to the castaway.

Above: An ARC Bell 412 operates over Antarctica. Since 2015, the force sends one of its offshore patrol vessels (OPV) with a Bell 412 helicopter to Antarctica every year. (Photo: Colombian Navy)

Left: The Fennec performs a rescue demonstration in the waters of Base Naval de Cartagena. (Photo: Colombian Navy)

One of the two AS365N Dauphins purchased in 2018. This is the example bought for the DIMAR (Dirección General Marítima, General Maritime Directorate) for search and rescue (SAR) operations, seen on its first tests on the deck of the Almirante Padilla corvettes. (Photo: Colombian Navy)

Above: The four Bell 412EPs of Grupo Aeronaval del Caribe (GANCA) received Telephonics APS-143C V3 synthetic aperture radar on the belly, for surveillance and SAR missions. (Photo: Colombian Navy)

Right: One of the four Bell 412EPs of GANCA. The radome of the radar is seen on the belly. (Photo: Colombian Navy)

Below: One of the two Dauphins, seen on the facilities of the Policía Nacional de Colombia (PNC) at Guaymaral Airport. (Photo: Colombian Navy)

GANPA

Based at Juanchaco, this group was created in 1997 to increase the presence of the force on the Pacific coast. This coast has the difficulty of having many swamps, islands, and small rivers, making it a prime spot for drug traffickers to hide their small boats or submersibles before they head out to sea. As the Cauca Valley, in the southwest of the country, is one of the main production areas for drugs, drug traffickers are very active on the coasts close to it.

Because of its location, in an area of swamps and forests, the Juanchaco base is only accessible by air or by water, with no roads to reach it; however, it is only 6km from NB Bahía Málaga, the main base of the ARC in the Pacific coast.

The unit is also divided into two squadrons, one of fixed-wing and one of rotary-wing, although they are smaller than those of GANCA. Currently, the first has one Grand Caravan and one Cessna 206, while the second operates three Bell 412s. Usually, one of the Persuaders is deployed at Juanchaco, but it belongs to GANCA. The GANPA also operates two Scan Eagle UAV systems.

Left: A GAU-17 and a Barrett 1.27mm rifle installed on the door of a Bell CUH-1N. The Barretts are used against the engines of the 'go-fast' boats.

Below: The CUH-1Ns are used by GANCA, Grupo Aeronaval del Pacífico (GANPA) and Grupo Aeronaval Central (GANCE) for troop and cargo transport and have seen a lot of action against drug traffickers and guerrilla groups.

A CUH-1N performs a rescue on the sea, using its hoist.

One of the Bell 412EPs before receiving its radar on the belly.

GANCE

The Grupo Aeronaval Central is based at El Dorado airport in Bogotá, but also deploys to other airports to the south and east of Colombia. Currently, they have the ATR 42, two Grand Caravans, one Cessna 206, five UH-1Ns, a single Bell 212, two Beechcraft C90s, a 350i Super King Air and a 350ER Super King Air.

As El Dorado airport is extremely busy with international and domestic passenger, cargo, military and police flights, the operation of slow aircraft has been restricted at certain times, so the Comando de Aviación Naval is planning to move one Grand Caravan to the east and the other to Puerto Leguízamo in the south. The Cessna 206 still operates from Puerto Carreño. The UH-1Ns are deployed at Puerto Leguizamo, on the border with Ecuador, in the Putumayo Department, and at Tumaco, Nariño Department, in the southwest end of the country. At the latter, the helicopters support the Fuerza de Tareas 72 Contra el Narcotráfico (Task Force 72 Against Drug Traffic). The UH-1Ns also operate in support of the marine forces and take part in combat regularly. On occasion, they operate from the deck of the eight PAFs (Patrullera de Apoyo Fluvial, Riverine Support Patrol Vessel) built locally by Cotecmar shipyard and equipped with a flight deck.

The ARC 212 is the only Bell 212 currently in service with the Navy, used by the GANCE on operations in the south and east of the country.

One of the CUH-1Ns flies over a Griffon hovercraft on the Putumayo River. This river is on the border with Ecuador and Peru, where there is a lot of drug trafficking and guerrila activity. (Photo: Colombian Navy)

Above: The ATR 42 provides an important capacity for passenger transport. (Javier Macías)

Right: The new ARC101, a craft 350i Super King Air delivered in 2011, which replaced a Beech 350 with the same serial that retired in 1999. (Photo courtesy of Beechcraft)

A UH-1N approaches one of the Patrullera de Apoyo Fluvial (PAF)-type riverine patrol vessels, which are the only ones equipped with a helipad. The boats are armoured and heavily armed to fight against the guerrillas. (Photo: Colombian Navy)

Chapter 3
Aviación del Ejército Colombiano – Colombian Army Aviation

Until 1995, the EJC depended on the FAC for all of its air support, but, after fighting for a long time to have its own air arm, on 25 August 1995, President Ernesto Samper authorised the creation of the Brigada de Aviación del Ejército (BAVE, Army Aviation Brigade), which was organised during 1996 and activated in August 1997. The EJC first focused on organising a helicopter force, for air assault and transport, purchasing ten Mil Mi-17-1Vs and seven UH-60L Black Hawk helicopters, with which two battalions were created: the Batallón de Aviación No 1 Carga Aérea (No 1 Aviation Battalion Air Cargo) and Batallón de Aviación No 2 Asalto Aéreo (No 2 Aviation Battalion Air Assault). At the same time, the Brigada de Aviación 25, dependant of the 5º División de Ejército (5th Army Division), was organised. Later, the organisation changed, as the force grew, until reaching the current size and deployment.

The force still has helicopters as its main asset, with a single battalion of fixed-wing aircraft and four of helicopters, one for each model, being the Black Hawk, Mi-17, CUH-1N and Huey II. Originally, the fixed-wing force was formed of impounded aircraft taken from drug traffickers, some of them captured before the Army's aviation force was officially created, but these are being replaced by new aircraft more suited to its mission. In 1997, the main Army Aviation base was created at Fuerte Militar Tolemaida (Tolemaida Military Fort), a large EJC base at Melgar, where there was a small runway used for paratrooper training. The facilities have been enlarged throughout the years, with a longer runway and more hangars.

War and threats
Despite the peace treaty with the FARC guerrillas, the war in Colombia is far from over. Large groups of FARC dissidents have not accepted the treaty and continue their combat operations. Furthermore, the ELN continues to grow, welcoming many disenfranchised men from the FARC and gaining control of the bulk of drug production and illegal mining. These guerrilla groups have also mutated from an organisation in fronts, with large concentrations of force of up to 800 men, to small groups of 15 to 20 men, without uniform. They try to avoid contact with the EJC and just make terrorist attacks and then escape, making them more difficult to fight.

In recent years, tensions with Venezuela have increased and led to a change from the counterinsurgency doctrine of the Fuerzas Militares de Colombia to a new strategy against a possible conventional conflict. The Venezuelan government is also helping the guerrilla groups, as forces of the ELN and FARC dissidence operate in Venezuelan territory without challenge.

Future plans
As for the future, one of the main plans of the EJC is to buy about ten small transport aircraft, with 2 tons of capacity, to replace the three CASA C-212-100s and increase their transport capabilities.

Among the models studied is the Dornier 228 and the PZL M-28. Furthermore, as previously mentioned, there is a programme, together with the other forces, to replace the UH-1H and CUH-1N fleets with a single model, with the Leonardo AW139 being preferred, but they are also analysing the Bell 412 and other types.

Current organisation

After some changes through the years, aviation within the EJC is currently organised as part of the División Asalto Aéreo (Air Assault Division), which has five brigades, three of them comprising aviation and the others of ground forces. All training of the crews is currently performed by the FAC, after they are selected from the other arms of the EJC.

Brigada de Aviación No 25 de Misiones de Aviación (Aviation Brigade No 25 of Aviation Missions)
This is the oldest unit of the EJC's aviation wing and is the only one equipped with its own aircraft. It has six battalions with aircraft, called Batallón de Aviación (BAAV, Aviation Battalion), the Batallón de Entrenamiento y Reentrenamiento de Aviación (BETRA, Aviation Training and Retraining Battalion) and the Batallón de Operaciones Especiales de Aviación (BAOEA, Aviation Special Operations Battalion). All of them have their own aircraft.

BAAV 1
BAAV 1 is based at El Dorado airport and comprises the fixed-wing aircraft, divided into four companies. Compañía Alfa (Alpha Company) has eight Grand Caravans, received since 2006, three of them being the EX version (received in 2013, 2015 and 2017, respectively), and the first five were modernised with Garmin 1000 avionics. They are mostly assigned to operate for command and control and communications with the ground troops, and also for medical evacuations. Two of them are equipped with door-to-drop paratroopers, carrying up to ten of them. They are at Yopal (VIII División de Ejército), Apiay (IV División), Medellin (VII División) and Cali (III División, Comando Conjunto Pacífico, Pacific Joint Command), while one is at Tolemaida to support the helicopters and to train paratroopers, with crews rotated every two weeks. The other three are usually at El Dorado on maintenance.

Compañía Bravo is equipped with aircraft of the King Air family. It has one C-90, which is used to support the helicopter fleet and sometimes for command and control. There are also three B200 Super King Airs and one B200 Catpass for liaison and intelligence, with one B200 equipped for ELINT missions with UHF and VHF and equipment to intercept cell phone communications, and the Catpass and one B200 equipped for intelligence, with vertical cameras and FLIR, as well as the UHF and VHF

Below left: One of the three Grand Caravan EXs pressed into service in 2013.

Below right: One of the Grand Caravans of Batallón de Aviación 1 (BAAV, Aviation Battalion).

Above left: This aircraft, serial number EJC 1138, operates mainly from Tolemaida in support of special forces training.

Above right: A Grand Caravan prepares for a flight at El Dorado facilities. The saturation of flights at the airport means there is pressure to move the operations of slow aircraft to another airfield.

equipment. Since 2010, they have been modernised with Garmin 1000 avionics. There are four B350i Super King Airs, three of them delivered between 2009 and 2010 and the last one in 2020. Of the first three, one is used for intelligence, one for command and control with satellite communications for the Comandante en Jefe del Ejército (Army Commander-in-Chief) and one for medical evacuation. They have Collins Proline 21 avionics. The last one was Beechcraft's demonstrator and is used as VIP transport, wearing the colours of the factory.

Compañía Charlie specialises in cargo missions, and it is equipped with three CASA C-212-100s, one Antonov An-32A and one An-32B. The first two CASA C-212s were delivered in 2006 by the Ejército del Aire, and they were modernised and delivered again in 2015 with new avionics and improved interior, while the third was transferred from the ARC in 2017. The An-32s were received in 2008 and 2009, respectively, impounded from drug traffickers, having the capacity to carry 35 to 40 soldiers, 50 passengers or 6.7 tons of cargo. In 2014, the An-32B, serial EJC-1147, was modified on the cabin floor to perform air drops of cargo, and in 2015, they performed the first drops, carrying up to 6 tons on six pallets, at a speed of 150 knots and between 800ft and 900ft of height. The first operational drop, with 5 tons, was in January 2015 in support of the Brigada Móvil 22 (Mobile Brigade 22) of the Fuerza de Tareas Conjunta Omega (Joint Task Force Omega), at Peñas Coloradas, Caquetá, a place that can only be reached by river or air, but in the winter season the river is almost dry. They usually operate on unpaved runways with up to 5 tons of cargo, and they are used to carrying up to 1,500 gallons of fuel drums to San José del Guaviare, on the Amazon, to support the troops there and provide medical evacuations. With special troops, they perform HAHO (high altitude, high opening) and HALO (high altitude, low opening) operations, both dropping them from 35,000ft and the first with an aperture over 6,000ft and the other below 600ft.

Compañía Delta has one Aero Commander RC690D, one RC695A, one RC690B and one RC695 used for transport, command and control, support, training and medical evacuation. The RC690B and RC695 were donated in 2005, and the other two were received from the FAC in 2013 and 2014, respectively. They were all modernised by the end of 2014 with the Renaissance Commander Program, one receiving Garmin 950 avionics and the others Garmin 650. There are plans to acquire more.

Aviación del Ejército Colombiano – Colombian Army Aviation

Above left: One of the 350i Super King Airs used for electronic intelligence missions.

Above right: The Ejército Nacional de Colombia's (EJC, National Army of Colombia) single Beechcraft B200 Catpass, equipped for surveillance with an FLIR turret.

Right: EJC 1125 is a 350i Super King Air used for command and control.

Below: EJC 1120 is another 350i, used for electronic intelligence and surveillance.

Above left: This Beech 350i, serial number EJC 1125, departs from the El Dorado facilities of the BAAV-1. Also at this airport is the Comando de Aviación de Ejército (Army Aviation Command).

Above right: One of the B200 Super King Airs used by BAAV 1.

Left: Two Antonov An-32s were impounded from drug traffickers. This is the EJC's only An-32A.

Below: One of the three CASA C-212-100s of BAAV 1.

Above: The EJC has three C-212-100s that once belonged to the FAC. An additional one was transferred from the ARC in 2017.

Right: The An-32B was modified to drop cargo by parachute in 2015.

Below: The second CASA C-212-100 being modernised at the maintenance facilities at El Dorado in 2015.

Above: One of the An-32s takes off from Tolemaida, while the other and a CASA C-212 are on the platform, alongside some Mil Mi-17 helicopters.

Left: An Aero Commander RC690D taxies at Tolemaida. These are mainly used for liaison missions.

Below: The EJC has four Aero Commanders of different models for light transport and intelligence.

BAAV 2

BAAV 2, based at Tolemaida, provides air assault assets for the ground forces, but its aircraft are also used for support and transport missions and to support the community in case of disasters. It is equipped with 42 S-70As and UH-60L Black Hawks, with the first seven UH-60Ls arriving in 1997, followed in 2001 by seven S-70A-41s and 14 more as part of the Plan Colombia. In 2005, three more UH-60Ls were purchased as part of Plan Escudo, and, in 2006, eight built for, but never delivered to, the Ejército de Venezuela (Venezuelan Army) were sold to the EJC. In 2009, 15 UH-60Ls were purchased from US Army stocks.

These aircraft carry out air assaults with regular and special forces and are used with a crew of two pilots, one mechanic/gunner and one gunner, and they are usually armed with M-60 or GAU-17 machine guns. The GAU-19 was tested, but its weight and the space used by the ammunition limited the transport capacity of the helicopter. For special operations, when the unit is attempting to demobilise guerrilla forces, such as in June 2011, the aircraft were equipped with six LRAD 1000 very-long-range acoustic devices with which they issue demobilization messages, and the high noise has the capacity to stress the enemy troops for a long period after hearing them. They can be used up to 5,000ft high.

For special forces in the jungle, the special patrol infiltration/exfiltration system (SPIES) is widely used. This system is also used for medical evacuation in the jungle. For regular assault operations, these aircraft carry up to 18 soldiers with their equipment and usually operate at night; all helicopter crews are trained to use NVGs, since most assault and attack missions are carried out at night. The Black Hawks received M-130 chaff/flare dispensers and ALQ-144 IR jammers.

Right: An UH-60L Black Hawk flying close to Tolemaida.

Below: A Black Hawk deploys troops of the Escuela de Lanceros (Lancers School) onto a lagoon. They are one of the elite units of the EJC.

Above: Troops of the Escuela de Lanceros rappelling from a Black Hawk of BAAV 2.

Left: A Black Hawk lands to pick up personnel of the Escuela de Lanceros. Sixty-one Black Hawks of the variants UH-60L, S-70A-41 and Sikorsky S-70i were delivered in batches to the force over a period of more than 20 years.

Below: A line of Black Hawks at Tolemaida, the main Aeródromo Militar base in Colombia.

Above: An S-70A-41 Black Hawk just landed. The M240 machine gun is visible on the window.

Right: One of the LRAD 1000 very long-range acoustic devices installed on Black Hawks to send messages to demobilise the guerrillas or to stress the enemy forces.

Below: An S-70A-41 Black Hawk of BAAV 1 lands on the training ground at Tolemaida.

Above: The UH-60L Black Hawk of the EJC can be recognised by its radar.

Left: A UH-60L with the LRAD 1000 very long-range acoustic device installed on the door.

A Black Hawk taxies are the sun sets at Tolemaida. There are always four members of crew on board.

BAAV 3

BAAV 3, also based at Tolemaida, received eight Mi-17-1Vs, six Mi-17MDs and eight Mi-17-V5s, the last two versions equipped with a hydraulic-operated ramp on the rear and other improvements, especially in avionics and engine power, which give a better performance in hot and high operations. These perform combat support and transport operations, as well as humanitarian missions. They are mainly used in the jungle, when they have to fly long distances to carry supplies to advanced posts or deployed troops, thanks to the three hours of endurance with internal fuel or five hours with extra fuel tanks. One Mi-17-1V, EJC-3375, used on the Operation *Jaque* (*Check*) to free 15 men hijacked by guerrillas in 2008, is now used for humanitarian missions, wearing a white and red livery. All Mi-17s received the ASO-2V countermeasure systems and the L-166V-1E Ispanka IR jammers. Some time ago, two aircraft were tested with GAU-19 and seven-tube rocket launchers and called 'Depredador' ('Predator') but this led to a conflict with the FAC, which has the responsibility for the attack operations, and the project was cancelled.

Right: An Mi-17-1V of BAAV 2 flies low over Tolemaida.

Below: The Mi-17s are mainly used for heavy transport and support, but also for humanitarian missions and firefighting.

Above left: An Mi-17-1V during a training mission close to Tolemaida. These are the heaviest helicopters in the Fuerzas Militares de Colombia (Military Forces of Colombia).

Above right: Although the model 'Mi-17-1V' is painted on the fuselage, EJC3388 is an Mi-17MD. Here, it is being readied for a firefighting mission with a 2,500-litre Bambi Bucket.

Currently, a total of 21 aircraft are on strength, including EJC-3378, which was received to replace one with the same serial, lost in an accident. This helicopter arrived with new avionics, including a new GPS, new radios and two displays, serving as a basis to start the modernisation of the others, with EJC-3398 being the first to receive the equipment when it was sent to Russia for overhaul. It was followed by a two more between 2018 and 2021, and plans are to continue in 2022 with the modernisation of others. In 2018, the force added the capacity to perform 1,500-hours inspections, with the support of the Russians, at Tolemaida. The first example overhauled in Colombia was EJC-3386, returning to service in early 2019. This not only reduces the cost of the overhauls, but also the time from nine to six months. At 4,000ft, the helicopters have a maximum take-off weight of 13 tons. However, at 12,000ft, the Mi-17V5s can take off with only 600kg to 700kg, as they are heavier than the Mi-17-1Vs because of extra armour. For firefighting, these aircraft use 2,500-litre Bambi Buckets.

EJC3375 is the helicopter used on Operation *Jaque* (*Check*) to free hostages, and it now wears this special scheme, based on the one used during the operation. It is now mainly used for humanitarian missions.

BAAV 4

This unit concentrates all of the CUH-1Ns, 33 of which were delivered by the US government through Plan Colombia, purchased second hand from the Canadian Armed Forces (where they were called CH-135); six of these were later transferred to the ARC and four to the PNC. Currently, this unit has 14 helicopters (two were lost in accidents, one is used for ground training and six were retired) being used for transport, assault and, under the name 'Cazadores' ('Hunters'), providing security to the assault helicopters, armed with two GAU-17 machine guns with 4,500 rounds per gun. For night operations, in 2011, they tested the Elbit Elop Mars and Lily-L night sights for the door gunners, both with an attached ESC Baz Max display, which significantly improves aiming capabilities; however, these were not adopted in the end.

These aircraft also act as command and control for the Infantry Division commanders, installing the radio of the commander in the rear cabin, or they perform medevac missions. For those missions, as for transport, medevac and assault ones, they are usually armed with M-60 machine guns instead of the GAU-17. The aircraft have had HALO capacity added to drop special forces and are developing the jump with static line. Furthermore, the EJC started to use the helicopters for TEPAM missions (Tirador de Élite de Plataforma Aérea Móvil, Mobile Air Platform Elite Shooter), and added thermal sights for the Barrett rifles that are used.

The unit also performs rescue missions with a hoist, even during the night, as their pilots have an average experience of 500 hours flying with NVGs, and some reach more than 1,300 hours. Fifteen CUH-1Ns were modernised with new avionics (Garmin 1000) and engines, and, since 2012, the airframes have been completely overhauled to extend the service life for at least ten years more. However, there is now a programme to replace these and the Huey IIs with a new model.

One of the CUH-1Ns that belonged to the Canadian Air Force. It was transferred to Colombia by the US Department of State Air Wing.

Above left: A CUH-1N takes off, loaded with troops, for a patrol close to Tolemaida.

Above right: A CUH-1N of BAAV 4 lands to pick up troops at Tolemaida.

Above: The CUH-1Ns are used for assault missions, light transport and escort duties.

Left: The EJC is now aiming to replace its fleet of CUH-1Ns, as well as one of the Huey IIs, with a new type.

Above: Currently BAAV 4 has 14 CUH-1Ns, which are expected to serve for about a decade more.

Right: A CUH-1N lands at Tolemaida. The base is about 70km to the southwest of Bogotá, next to BA Melgar of the FAC.

Below left: A CUH-1N, armed with a GAU-17 and equipped with an extra fuel tank. Some CUH-1Ns received the FastFin.

Below right: A line of CUH-1Ns and a Black Hawk at Tolemaida.

BAAV 5

The last battalion is BAAV 5, equipped with the remaining 22 examples of 32 Huey IIs delivered in 2002 as part of the Plan Colombia, and 2005 as part of the Plan Escudo. They are used for the same missions as the CUH-1Ns, but they started the TEPAM training before, in 2016. These operations are performed with the Comando Especial Contra Amenazas Transnacionales (CECAT, Special Command Against Transnational Threats), against high value targets. Usually, they include a shooter with a Barrett rifle and an observer. The aircraft of this unit also started paratrooper operations before BAAV 4. Like the CUH-1Ns, the Huey IIs always fly with extra fuel tanks on the cargo cabin to have 3.5 hours of endurance and all have extra armour on the floor. For rescue operations, they received a hoist in the cabin too.

Since 2015, the aircraft have started to receive new avionics and communications equipment, with EJC-5419 being the first modernised, and currently two are operational, after being converted by the Helicentro company in Bogotá. The others were modernised by the EJC at Tolemaida. In February 2013, a new Huey II was received with an improved engine exhaust, an FLIR Star SAFIRE III turret, a search light, and a radar, for use as command post, with a moving map and a station on the back for the operator of the FLIR, as well as a display on the co-pilot's panel. For firefighting, they use 420-gallon Bambi Buckets.

EJC5433 is a Huey II that was received in 2013, already equipped with an FLIR Star SAFIRE III turret, a Trakka Beam search light and a radar, as well as a longer engine exhaust, for use as command post. (Photo: Colombian Army)

Above: The Huey IIs of the EJC belonged to the US Department of State Air Wing before they were transferred to Colombia.

Right: A Huey II at the auxiliary runway of Tolemaida.

Below left: The EJC aims to replace their fleet of Huey IIs as soon as possible.

Below right: A total of 32 Huey IIs were delivered to the EJC between 2002 and 2005, of which 22 still remain operational.

BETRA

This unit is dedicated to the specific training of the crews on each type of aircraft, as well as in all different tasks, preparing pilots from FAC aviation schools to operate the way the EJC does. As the constant operations made it difficult to have helicopters constantly available for training, in early 2017, the Pelotón Puerta Naranja (Orange Door Platoon) was organised, depending directly on the Brigada 25, with one helicopter of each type destined for training missions, all equipped with orange doors. Furthermore, training areas were organised in Tolemaida for confined area operations on mountains, lagoons, etc. For operations on high mountains, they usually deploy to Tunja, Boyacá Department, to the north of Bogotá, with the Mi-17s and UH-60L Black Hawks, to perform training up to 13,000ft. For those operations, they mainly use the Mi-17 used in Operation *Jaque*.

Above left: We can tell that these helicopters of the Batallón de Entrenamiento y Reentrenamiento de Aviación (BETRA, Aviation Training and Retraining Battalion) belong to the Pelotón Puerta Naranja (Orange Door Platoon), thanks to their orange doors.

Above right: BETRA operates a single example of each helicopter type for training.

Below: An orange-doored CUH-1N flies close to Tolemaida during a training sortie.

The Batallón de Operaciones Especiales de Aviación (BAOEA, Aviation Special Operations Battalion) operates the fleet of Sikorsky S-70is, such as the one in this image, and also some UH-60Ls.

BAOEA

Created in 2009, this unit depends directly on the Bogotá-based División Asalto Aéreo, although it is based at Tolemaida. The unit received seven S-70i Black Hawks, five of them in January 2013 and the others in July. These are the most modern versions of the helicopter, with MFD, GPS/INS and digital maps, among other equipment, to aid in navigation and operations involving rescue hoists. The unit also has five UH-60Ls.

The battalion was based on the organisation and operation of the 160th Special Operations Aviation Regiment (Airborne) of the US Army, and performs operations against high-value targets, mainly the commanders of the FARC and ELN guerrillas. They also have an SAR company, trained for night operations, with the hoist equipped with an infrared light for use with NVGs, to avoid being detected by the enemy. They also perform rescue operations in case of disasters and were deployed to Haiti after the earthquake in 2010, and to Peru in 2017 after the floods. The unit also has a company that operates nine RQ-11B Raven drones for tactical missions, received in 2013, which are used to detect the operation of the guerrillas or other illegal activities. Currently, it has teams deployed to the Alta Guajira, Cúcuta and Tibú, close to the border with Venezuela and an area with a large presence of ELN guerrillas.

The other brigades

The Brigada de Aviación No 32 de Apoyo y Sostenimiento (Aviation Brigade No 32 for Support and Sustainment) has its headquarters at Tolemaida, and it is in charge of maintenance. The Batallón de Especialistas de Mantenimiento de Aviación (BAEMA, Aviation Maintenance Specialist Battalion) are specialists in equipment and comprise the Batallón de Mantenimiento de Aviación 1 (BAMAV 1, Aviation Maintenance Battalion), the only maintenance unit located at El Dorado, in charge of the maintenance of the fixed-wing aircraft, the BAMAV 2 for the maintenance of the UH-60, the BAMAV 3 for the Mi-17, the BAMAV 4 for the CUH-1N and the Huey II and the Batallón de Apoyo, Abastecimientos y Servicios para la Aviación (BAAAS, Aviation Support, Supply and Services Battalion), which is in charge of the logistics, supplying spares to all the maintenance units.

The Brigada de Aviación No 33 de Movilidad y Maniobra (Aviation Brigade No 33 of Mobility and Maneuver) is an operational unit, with eight battalions. While it does not have its own aircraft,

Above left: The maintenance facilities of Batallón de Mantenimiento de Aviación 2 (BAMAV 2, Aviation Maintenance Battalion) at Tolemaida, used for the Black Hawk fleet.

Above right: BAMAV 3 is in charge of the maintenance of the Mi-17 fleet.

it receives helicopters from BAAV 2–5 to perform its combat operations. The battalions are based at Santa Marta, Bucaramanga, Popayán, Apiay, Neiva, Larandia, Medellín and Saravena.

The Brigada de Fuerzas Especiales Rurales (Rural Special Forces Brigade) are ground troops of special forces used as a fast deployment force against the guerrillas, while the Brigada Especial Contra el Narcotráfico (Special Brigade Against Drug Trafficking) is a special unit against drug traffickers. Both are dependant to the División Asalto Aéreo. El Departamento de Alistamiento para el Combate y la Seguridad Aérea (Air Safety and Combat Readiness Department) oversees the safety and maintenance of the bases.

A view of Tolemaida airfield.

Chapter 4
Policía Nacional de Colombia – Colombian National Police

In a country marked by a strong presence of drug trafficking and communist guerrillas, the PNC has garnered the biggest aviation force of its kind around the world, equipped with more than 60 aircraft and 80 helicopters, including Huey IIs, Bell 212s and Black Hawks, all armed with machine guns. The force is dependant to the Ministerio de Defensa Nacional and is the only police force in Colombia, covering the entire territory of the country.

The aviation arm of the PNC was born in the 1950s, when the police of the Bolivar Department added a Ryan Navion for use on police duties. However, officially, the first aircraft of the force was a single Cessna 410 received in the early 1960s from the national railroad company, operated by the Director General de la Policía Nacional (National Police General Director), Mayor General (Major General) Bernardo Camacho Leiva. The aircraft was only used for a short time by Camacho Leiva, and an official aviation command was never organised. However, shortly after, in 1965, Mayor General Saulo Gil Ramírez Sendoya ordered the creation of a small aviation force, and one Cessna 441 and two Cessna 206s were added, with Tenientes José Francisco de Paul Loiseau Loiseau, Luis Gilberto Santamaría Arroyave and Humberto Aparicio Navia being the first pilots, after receiving training at the Aeroclub de Colombia (Colombian Air Club) and the Escuela Nacional de Aviación Civil de Colombia (National School of Civil Aviation of Colombia). Maintenance was performed at private workshops, flying from Guaymaral civil airport to the north of Bogotá and renting part of a civil hangar.

As the intention was to increase the aviation unit, in 1973 a unit called Sección de Transporte Aéreo (Air Transport Section) was created, the name of which changed to Grupo de Transporte Aéreo in 1979. In the meantime, two extra Cessna 206s were received. The first helicopters were two Bell 206L-1s and one Bell 212, donated by the US government in 1981 to fight against the increasing drug trafficking and the Policía force that had been tasked to be part of the war against it. As the force did

Below left: The Basler BT-67s are still among the main transport aircraft of the PNC.

Below right: Currently, the PNC has two 208B Grand Caravans. The example in the picture has since been sold to a private operator.

not have the infrastructure to operate the helicopters, they were given to the nation's Attorney General, but one year later they were returned to the PNC, which started to prepare its own pilots.

On 29 July 1983, by Presidential Decree 2137, the Servicio Aéreo de la Policía (Police Air Service) was created, dependant to the Dirección Operacional (Operational Direction). On 1 November, the first course of officer pilots started at the Aerocentro de Colombia Flying School, and, after receiving the initial training, on 4 February 1984, they continued training with the FAC, together with the first mechanics. Meanwhile, the first base of the force started to be built at Guaymaral, with US support, which was inaugurated on 10 November 1985 under the name of BA Guaymaral Capitán Fernando Álvarez Bonilla. Furthermore, the construction of bases at Santa Marta, San José del Guaviare and Valledupar began.

Between 1983 and 1985, the fleet was expanded with three new Bell 212s, three Bell 206B-3s, one DHC-6-200 Twin Otter and three Ayres Turbo Thrushes. The last aircraft and the Bell 206s were used to spray the coca fields and eradicate them, being deployed to the jungle airfields with the support of the Bell 212s. The Bell 206s received a spraying device on the belly, with a tank and a boom. However, the increase of the drug trafficking and production activities showed the need to augment the strength of the Policía's small aviation division.

In 1986, the US government donated one Bell 212, six Bell 206L3s, a second DHC-6 Twin Otter, one Beechcraft C-99 (impounded from drug traffickers) and a single Cessna 152. The Colombian government purchased one extra 212, four 206L3s and a second 152. All this new equipment arrived between 1986 and 1987. With the two Cessnas and two of the Bell 206s, the PNC started to train their own pilots, creating the Escuela de Aviación Policial (Police Aviation School). In 1987, the first women were admitted for pilot training, with Teniente Luz Nancy Parrado Amaya being the first woman to gain her wings.

On 3 March 1987, by Presidential Decree 423, the Servicio Aéreo de la Policía became part of the Dirección Antinarcóticos (Anti-narcotics Directorate), together with the Servicio Especializado de Control de Sustancias que producen Adicción Física o Síquica (COSAS, Specialised Service to Control the Substances That Produce Psychic or Physical Addiction). Also in that year, land was given to the force at El Dorado International Airport in Bogotá to install a base for transport aircraft. The facilities were inaugurated in 1989 with the two Twin Otters, the Beechcraft C-99, one Piper Conquest II, one Rockwell Turbo Commander and a Piper Navajo, followed later by two Beechcraft 200s and one 300 Super King Air assigned by the Ministerio de Justicia (Ministry of Justice).

In 1989, the US aid continued with 12 UH-1Hs, two Hughes 369Ds (500D) and one 500E, followed shortly after by two 208 Grand Caravans, one Piper Seneca II, two additional Turbo Thrushes, six Basler BT-67s and six Cessna U-206s. In 1994, one Bell 206B-3, four Bell 212s and one Super King Air were received, while deliveries of UH-1Hs continued until the force had 56 examples, as well as Bell 212s, which reached a total of 19 examples. During the 1990s, a single Fairchild C-123 Provider captured from drug traffickers was pressed into service, but it was only used briefly as the aircraft was old and difficult to keep operational. In February 1996, the Escuela de Aviación Policial was moved to new facilities inaugurated at Mariquita, Tolima Department, also close to Bogotá, with two Cessna 152s, all the Cessna 206s and some Bell 206 helicopters. Training of new pilots began at that base in 1997.

In 1999, through an agreement with the US government, 11 UH-60L Black Hawks, one Bell 412EP and 24 Huey IIs were received, while 12 UH-1Hs were taken to that standard (another 12 were lost in accidents or in combat, and the other 32 were retired). The fleet was increased again in the new millennium, with three Beechcraft B-200s, one ATR 42, two Cessna 172s, three Cessna 206s, three Piper Navajos, one Piper Cheyenne, 12 Bell CH-135s and seven Huey IIs from US stocks and six from the EJC. The CH-135s were retired before 2010, with four going to the EJC, two having been lost in accidents and six returned to the US.

One of the two Beechcraft 1900Cs used for transport.

In 2010, the first Bell 407 arrived, with the aim of replacing the Hughes 500 and the losses of the Bell 206L fleet. Between then and 2020, five 407s, one 407GX and one GXi were delivered, some of them equipped with FLIR Star SAFIRE 380HDC turrets with TV and thermal cameras, laser rangefinder, and also a Trakka Beam searchlight, which can be attached to the turret. The system can geolocate what the camera is seeing and integrate it with a map, determining the address of a place that is being seen with the camera.

In 2017, the fleet of Black Hawks was reinforced with ten UH-60As donated by the US government. In May 2021, three UH-1Hs arrived, similarly given by the US government. They had belonged to the Pakistani Air Force and had not yet been taken to Huey II standard. They are yet to be pressed into service yet. During the 1990s, the US DoSAW operated in Colombia with four Alenia C-27s, some Grand Caravans, 15 S2R Turbo Thrushes and 18 Air Tractor AT-802s, all with PNC serials added to the US ones. The C-27s and Grand Caravans were used to support the operations of the other aircraft, which were equipped for spraying and received extra armour. In 2002, they were aided by 11 OV-10D Broncos modified for spraying, with a tank on the rear stowage and the boom under the wing. At least two Air Tractors, three Broncos and seven Turbo Thrushes were shot down on illicit plantations eradication, and the surviving Broncos returned to the US in 2008. Seven Turbo Thrushes are still in operation in the country, all transferred by the US government to the Policía, and in 2017 two AT-802Fs were added, totalling nine aircraft in service.

An Ayres S2R-T65 Turbo Thrush used for spraying coca fields. Currently, the PNC only uses seven Air Tractor AT-802s and two AT-802Fs to spray coca fields.

Today

In the present, the fixed-wing force comprises four BT-67s, two Twin Otters, two Caravans, three Grand Caravans, three Piper Navajos, two Beechcraft 1900Cs, two 1900Ds, five ATR 42s, two Dash 8-311s, four B200 Super King Airs, two 300 Super King Airs, one Beechcraft 350 Super King Airs, six Fairchild C-26 Metros (two of them for intelligence), nine Cessna TU206Gs, three Cessna 172s and one Cessna 150L. The Metros for intelligence have an FLIR, radio scanning and other electronic and signal intelligence equipment, and one of them also has a high-definition vertical camera. They are used to intercept signals, short wave radios, cellular phones and locate its position. They also act as command posts, because they have satellite communications equipment that is used to support ground operations. The other Metros can be used for transport and medical evacuation, as they have conversion kits to carry two patients in intensive care.

Above left: One of the two DHC-6-200 Twin Otters of the force, which were modernised in 2010 and 2011.

Above right: This 300 Super King Air is used by the force for VIP transport.

Below: The Fairchild C-26 Metros are used for transport, medical evacuation and intelligence missions.

Above left: The BT-67s proved to be excellent aircraft for operations in the Amazon region.

Above right: One of the five ATR 42s of the force, which are the biggest transports it has.

Above left: A Metro taxies at El Dorado Airport.

Above right: A Cessna TU206G, which is used for liaison. While these aircraft are maintained at Guaymaral, usually they are deployed to small units across the country.

The BT-67s are the main transport aircraft, but the arrival of the ATR 42 marked the beginning of the replacement of this type, which is considered very reliable and cheap to operate. The Beechcraft B200 and B300 are used for VIP flights, and they are joined by one belonging to the Produraduría de la Nación (Nation's Prosecution), operated with the Policía serial to carry arrested people. Regarding the helicopters, it has nine UH-60L Black Hawks, ten UH-60As, one Bell 412EP, eight Bell 212s, 32 Huey IIs, three UH-1Hs, six Bell 206Ls, six Bell 206B-3s, seven Bell 407s and two Hughes 500Ds.

Organisation

The Servicio Aéreo de la Policía is part of the Dirección Antinarcóticos, which falls under the Dirección de la Policía Nacional (National Police Directorate). It comprises five groups, which are: Grupo de Operaciones Aéreas (Air Operations Group), Grupo de Estandarización Aeronáutica (Aeronautical Standardisation Group), Grupo de Seguridad Integral (Comprehensive Security Group), Grupo de Mantenimiento Aeronáutico (Aeronautical Maintenance Group) and Grupo de

Abastecimiento Aeronáutico (Aeronautical Supply Group). The Grupo de Operaciones Aéreas is in charge of the operations and the aircraft and has five companies, as well as the Centro de Control (CEMAP, Control Centre).

Compañía Antinarcóticos de Aviación de Bogotá

This is the main unit, located at El Dorado airport, with the headquarters of the Servicio Aéreo de la Policía and the CEMAP, and it is from here that the aircraft are monitored via satellite. Each aircraft has a control panel, an Iridium antenna for satellite communications of voice and data, a GPS, audio and Sky Connect data processing boxes and Flight Explorer software. The control panel has a button to transmit when an emergency occurs, one for when the aircraft is receiving enemy fire, and one to indicate places of interest, of which the information is geolocated. The aircraft normally sends its location every three minutes, but when the buttons are pressed, it sends the information every 30 seconds. The software receives meteorological information from the command centre, which is then sent to the aircraft and the system serves to guide them in case any instruments are lost or damaged.

Compañía Antinarcóticos de Aviación de Guaymaral

This unit is located at Guaymaral, and it is the main helicopter and light aircraft unit, having three hangars at the airport where the major maintenance of the helicopters, Cessnas and Piper Navajos takes place. Most of the Huey IIs and Bell 212, 206, 407, the 412s and the Black Hawks, as well as the Piper Navajos and some of the Cessna 206s are based here.

Above left: A line of Huey IIs at Guaymaral. The PNC has 32 of them on strength.

Above right: A Bell 212 and 412 at Guaymaral Airport.

Left: PNC0733 is one of the few Huey IIs of the force to be equipped with radar. It was baptised *Excalibur*.

Above left: A Huey II flies to the north of Guaymaral during a test flight.

Above right: One of the Bell 212s at Guaymaral. Some are painted on white and green and others dark green.

Above: The Huey IIs are widely used on operations against drug traffickers.

Right: The Bell 206s are used for city patrol and training.

Above left: A Bell 212 in the dark green paint scheme.

Above right: A line of helicopters at Guaymaral, with the Bell 412, a 212 and some Black Hawks.

Above: One of the UH-60L Black Hawks received by the force. This example was lost in 2015 in an accident.

Left: One of the UH-60Ls equipped with radar, pictured during a flight to train a hoist operator.

Above left: Currently, the PNC operates a total of 19 Black Hawks of the models UH-60A and L.

Above right: The Black Hawks are mainly used on assault operations against drug traffickers, guerrilla groups and illegal mining.

Above: A crowded tarmac at Guaymaral, with eight Black Hawks of the PNC.

Right: A Bell 407GXi with FLIR Star SAFIRE 380HDC turret with TV and thermal cameras, laser rangefinder, and a Trakka Beam search light.

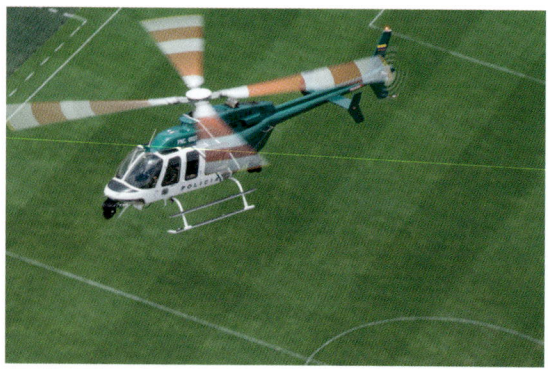

Above left: A Bell 407 of the PNC flies over El Campín, the national stadium, in Bogotá.

Above right: A Hughes 500 is seen at Villavicencio Airport. They are used for urban patrols.

Above left: While a Hughes 500 was tested with a GAU-17 minigun, this addition was scrapped, as the helicopter had little ammunition capacity.

Above right: The Bell 407 is a big step up from the older 206 and Hughes 500, thanks also to its cameras and search light.

Left: The PNC aims to add more Bell 407s to its fleet.

A Cessna 172N, which is used at Mariquita at the Escuela de Aviación Policial (Police Aviation School).

Escuela de Aviación Policial

The school is located at Mariquita and is dependant to the Grupo de Estandarización Aeronáutica. It uses a Bell 206B and L, a single Cessna 150 and the three 172s for basic training. This unit also has an aviation company; it is equipped with Huey II helicopters, but its strength varies according to need.

Other units

The other companies are at Santa Marta (Magdalena, equipped with Bell 212s) and Tulua (Cauca Valley, equipped with Huey IIs). There are also smaller units at San José del Guaviare, Villa Garzón (Putumayo) and Tumaco, which have no assigned equipment but receive helicopters from Guaymaral. There are also Hughes and Bell 206s operating in Cali, Cartagena, Villavicencio, Pereyra (in the coffee production region) and Medellín, depending to the commander of each zone.

All units, except those at Bogotá and the Escuela de Aviación Policial, use only helicopters, and each unit performs its own minor maintenance work, while more serious work takes place at Bogotá. The Cessnas, Piper Navajos and helicopters receive maintenance at Guaymaral, all under the control of the Grupo de Mantenimiento Aeronáutico. The Grupo de Abastecimiento Aeronáutico is in charge of the logistics, while the Grupo de Seguridad Integral is in charge of the security of the aviation facilities.

The needs of the PNC have derived from the war against drugs, alongside operations against the guerrilla forces of the FARC and ELN, and these needs have led to a change in the orientation from aviation for preventive operations (like other police aviation arms around the world) to one with more repressive power. Furthermore, because the operations took place against very well-equipped forces and in places with limited access, like jungles and mountains, they organised a force with a lot of firepower and an impressive transport capability.

Operations

The aircraft are only used for transport and logistic support, as well as the fumigation, performed with the Air Tractor AT-802s. In the past, they were flown by US-contracted pilots, but the PNC trained its own and now performs those missions. For the eradication of the plantations, the force also increased the availability of manual jobs, inviting farmers to find alternative, legal work.

The helicopters operate in four kinds of missions: fighting against drug trafficking, anti-guerrilla operations and illegal mining, as well as the usual police work. The Huey IIs, Black Hawks and Bell 212s perform escort missions for the fumigation aircraft in packs of four, three armed and one for SAR. The armed ones cover the aircraft if they are being shot at, and the SAR always stays far and high, waiting to enter into action in case an aircraft or helicopter is shot down. For these missions, they usually operate GAU-17 machine guns of six barrels and 7.62mm calibre.

PNC aircraft also perform interdiction operations, concentrated on the destruction of laboratories, illegal runways, and drug seizures. As these missions are usually located in inhospitable locations, the force usually arrives by air, with support from the intelligence aircraft. The illegal runways are destroyed with explosives, with three or four holes. This work is performed with the Grupo Jungla (Jungle), the special forces of the police, which is prepared to operate in the jungle. For interdiction operations, the force uses the Huey II, Black Hawk and Bell 212; the selection depends on the complexity of the job, the place and the number of people to be carried. To attack laboratories, the Black Hawk is usually the chosen aircraft. It is also the chosen aircraft for operations against illegal mining, in which it performs air assaults to fight against the guerrilla forces protecting the camps.

The missions against guerrilla groups are specific operations, often aiming to capture a chief of the terrorist group, and they almost always operate with the FAC or the EJC. According to the needs of the operation, each force has its specific tasks. The EJC verifies the intelligence information, the FAC performs the bombings, and the PNC arrives to secure the location and gain full control of it, verify what has taken place and begin the judicial process. In these missions, Black Hawks are the primary aircraft, while the Bell 212 or Huey II can bring support or act as a command post. Most missions take place in the night, so all Black Hawk and Bell 212 crews are enabled to fly with NVGs. Until 2009, the Huey IIs were also enabled, but one helicopter was lost after the pilot got disorientated during a night operation, and operations with NVGs ceased.

For the operations against important targets, air assaults are made with Black Hawks in the middle of the night, like the ones made against the 'Mono Jojoy' and 'Cuchillo', the leaders of the FARC and the leader of a drug cartel, respectively. Although the helicopters are usually armed with GAU-17 machine guns, the Servicio Aéreo de la Policía also has 12.7mm GAU-19 miniguns, which are mainly used in the Black Hawks, but also on the Bell 212s and Huey IIs. The weapons also comprise 12.7mm Browning M2 guns and 7.62mm M240s. The force also bought flare dispensers, as it had detected MANPADS (man-portable air-defence systems) in the hands of guerrilla groups.

The Bell 412 is only used for VIP flights, while the Hughes, Bell 206 and 407s are used for city patrolling. While the 407s use Star SAFIRE 380HDC turrets, the 206s were equipped with older Star SAFIRE types and deployed to smaller cities across the country. The Hughes do not use cameras, but in the past, one MD530F was tested with a single GAU-17 machine gun, commanded by the pilot, but the small ammunition capacity and short range of the helicopter determined that the idea was scrapped. Gradually, the 407 is becoming the main helicopter for city patrol, and the plan is that the force will continue adding more examples. A new mission added to the force is firefighting, for which it received Bambi Buckets to be used with the Black Hawks and Bell 212s.

Chapter 5
Orders of Battle

Fuerza Aérea Colombiana order of battle

Command	Group	Squadron	Equipment
CACOM-1	Grupo de Combate 11	Escuadrón de Combate 111 'Dardos'	IAI Kfir C12 and TC12
		Escuadrón de Combate Táctico 113 'Fantasma'	Basler AC-47T 'Fantasma', IAI 201 Arava, Cessna 208B Grand Caravan
		Escuadrón Defensa Aérea 114	Airbus ECN-235-100M 'Phobos', CN-235-100M
		Escuadrón de Combate 116	Beechcraft T-6C Texan
CACOM-2	Grupo de Combate 21	Escuadrón de Combate 211 'Grifos'	Embraer EMB-314 Super Tucano
		Escuadrón de Combate 212 'Tucanos'	Embraer EMB-312 Tucano
		Escuadrón de Combate Táctico 213	Sikorsky AH-60L Arpía, Cessna 208B Grand Caravan, CASA C-212-300, Basler AC-47T Fantasma, Schweizer SA 2-37B 'Vampiro'
		Escuadrón Defensa Aérea 214 'Fénix'	Cessna SR-560 Horus
		Escuadrón de Combate 217 'Quimera'	Elbit Systems Hermes 450 and Hermes 900
CACOM-3	Grupo de Combate 31	Escuadrón de Combate 311 'Dragones'	A-37B Dragonfly
		Escuadrón de Combate 312 'Drako'	Embraer EMB-314 Super Tucano
		Escuadrón de Combate Táctico 313	Cessna 208B Grand Caravan, Bell 212 Rapaz, Beechcraft C90GTx, Embraer EMB-110 Bandeirante
		Escuadrón Defensa Aérea 314	Cessna SR-560 Horus
CACOM-4	Grupo de Combate 41	Escuadrón de Combate 411 'Rapaz'	Bell 212 and CUH-1N (CH-135)
		Escuadrón de Asalto Aéreo 412	Bell UH-1H, Huey II, Cessna 208B Grand Caravan, Bell OH-13H
		Escuadrón de Ataque 413 'Escorpion'	Hughes 369HS and MD 500E

Command	Group	Squadron	Equipment
CACOM-4 (Cont'd)	Grupo de Combate 41 (Cont'd)	Escuela de Helicópteros de las Fuerzas Armadas 'Coronel Carlos Alberto Gutierrez'	Bell 206 and Bell TH-67
CACOM-5	Grupo de Combate 51	Escuadrón de Combate 511	Sikorsky AH-60L Arpía
		Escuadrón de Combate 512	Sikorsky AH-60L Arpía
		Escuadrón de Operaciones Especiales 513	Cessna 208B Grand Caravan, Sikorsky UH-60L Black Hawk and MH-60L Black Hawk
CACOM-6	Grupo de Combate 61	Escuadrón de Combate 611	Embraer EMB-314 Super Tucano
		Escuadrón de Combate Táctico 613 'Pelicano'	CASA C-212-300, Cessna 208B Grand Caravan, Schweizer SA 2-37B Vampiro, Basler AC-47T Fantasma, Bell Huey II, Bell 212 Rapaz, Boeing Insitu Scan Eagle UAV
CACOM-7	Grupo de Combate 71	Escuadrón de Combate Táctico 713	AC-47T Fantasma, AH-60L Arpía IV, Bell 212 Rapaz, Cessna 208B Grand Caravan, CASA C-212-300
Escuela Militar de Aviación	Grupo de Educación Aeroáutica	Escuadrón Preparatorio 711	Stemme S-10 VT, SZD-54-2 Perkoz, Cessna T-41D, Cessna 172S
		Escuadrón Básico 712	CIAC T-90C Calima, Stearman PT-17 Kaydet, Beechcraft T-34A Mentor
CATAM	Grupo de Transporte Aéreo 81	Escuadrón de Transporte 811	Boeing KC-767, 727-2B7, 727-2X3F, 737-4S3SF and 737-46BF, Lockheed C-130B/H Hercules, Airbus C295M, Beechcraft C90GTx, Beechcraft 350, Cessna 208B Grand Caravan
	Grupo de Vuelos Especiales 82	Escuadrón de Transporte Especial 821	Fokker F 28 Mk1000, F 28Mk3000C, Embraer ERJ135BJ, Bombardier Learjet 60, Leonardo AW139, Bell 412, Sikorsky UH-60L Black Hawk
	Grupo de Inteligencia Aérea 83	Escuadrón de Inteligencia Aérea 831	Beechcraft 350 Super King Air, Beechcraft 300 Super King Air, Aero Commander 695
	Servicio de Aeronavegación a Territorios Nacionales (SATENA)		ATR 42-512, ATR 42-500, ATR 42-600, Embraer ERJ135BJ
Grupo Aéreo del Amazonas		Escuadrón de Combate 401	CASA C-212-300, Cessna 208B Grand Caravan
Grupo Aéreo del Oriente		Escuadron de Combate Táctico 201	Bell 212 Rapaz, Bell Huey II, Cessna 182R and Scan Eagle UAVs
Grupo Aéreo del Caribe		Escuadrón de Combate 101	Beechcraft C90GTx.

Command	Group	Squadron	Equipment
	Grupo Aéreo del Casanare	Escuadrón de Combate 301	Embraer Emb-314 Super Tucano, Sikorsky AH-60L Arpía IV, Cessna 208B Grand Caravan and Scan Eagle UAVs
CAMAN	Grupo de Transporte Aéreo 91	Escuadrón de Transporte 911	CASA C-212-300
Corporación de la Industria Aeronáutica Colombiana (CIAC)			Aircraft factory and modernisation centre
Fuerza de Tareas Ares			Special forces
Grupo de Operaciones Especiales Aéreas (GROEA)		Escuadrón de Comandos Especiales Aéreos	Special forces

Aviación Naval Colombiana order of battle

Group	Squadron	Equipment
GANCA	Fixed Wing Squadron	Airbus CN-235-200 and CN-235-300MP Persuader, Cessna 206
	Rotary Wing Squadron	Bell 412EP, 412EPi, CUH-1N, Eurocopter AS365N Dauphin and AS555 Fennec
GANPA	Fixed Wing Squadron	Cessna 208B Grand Caravan, Cessna 206 and Scan Eagle UAVs
	Rotary Wing Squadron	Bell 412
GANCE	Fixed Wing Squadron	Cessna 208B Grand Caravan, Cessna 206, Beechcraft C90 and Beechcraft 350i and 350ER Super King Air
	Rotary Wing Squadron	Bell UH-1N

Aviación del Ejército Colombiano order of battle

Brigade	Battalion		Equipment
Brigada de Aviación No 25	BAAV 1	Compañía Alfa	Cessna 208B Grand Caravan
		Compañía Bravo	Beechcraft C-90 King Air, B200 Super King Air, B200 Catpass and B350i Super King Air
		Compañía Charlie	Casa C-212-100, Antonov An-32A and Antonov An-32B
		Compañía Delta	Aero Commander RC695A C650, one RC690B and one RC695
	BAAV 2		Sikorsky S-70A and UH-60L Black Hawk
	BAAV 3		MiL Mi-17-1V, Mi-17MD and Mi-17-V5

Brigade	Battalion		Equipment
Brigada de Aviación No 25 (Cont'd)	BAAV 4		CUH-1N (CH-135)
	BAAV 5		Bell Huey II
	BETRA		Sikorsky UH-60L Black Hawk, MiL Mi-17-1V, CUH-1N (CH-135) and Bell Huey II
	BAOEA	Compañía de Operaciones Especiales y Asalto Aéreo (Air Assault and Special Operations Company)	Sikorsky S-70i Black Hawk and Sikorsky UH-60L
		Compañía de Búsqueda y Rescate de Personal C-SAR (C-SAR Personnel Search and Rescue Company)	Ground forces
		Compañía de Sistemas Aéreos No Tripulados Para la Maniobra Terrestre (Unmanned Aerial Systems Company for Ground Maneuvering)	AeroVironment RQ-11B Raven
Brigada de Aviación No 32 de Apoyo y Sostenimiento	BAEMA		Equipment maintenance
	BAMAV 1		Fixed-wing maintenance
	BAMAV 2		Black Hawk maintenance
	BAMAV 3		Mi-17 maintenance
	BAMAV 4		CUH-1N and Huey II maintenance
	BAAAS		Logistics
Brigada de Aviación No 33 de Movilidad y Maniobra	Batallón Movilidad y Maniobra de Aviación (Aviation Mobility and Maneuver) No 1, Buenavista		Uses aircraft from Brigada de Aviación 25
	Batallón Movilidad y Maniobra de Aviación No 2, Bucaramanga		
	Batallón Movilidad y Maniobra de Aviación No 3, Popayán		
	Batallón Movilidad y Maniobra de Aviación No 4, San José del Guaviare		
	Batallón Movilidad y Maniobra de Aviación No 5, Neiva		
	Batallón Movilidad y Maniobra de Aviación No 6, Larandia		
	Batallón Movilidad y Maniobra de Aviación No 7, Medellín		
	Batallón Movilidad y Maniobra de Aviación No 8, Saravena		

Brigade	Battalion	Equipment
Brigada Especial Contra el Narcotráfico No 5	Batallón Contra el Narcotráfico No 1 'Brigadier General Rodolfo Herrera Luna'	Ground forces against drug trafficking
	Batallón Contra el Narcotráfico No 2 'Coyaimas'	
	Batallón Contra el Narcotráfico No 3 'Mayor Pedro Solaque Chitiva'	
	Batallón Contra el Narcotráfico No 4 'Mayor General Alfredo Bocanegra Navia'	
	Batallón de Apoyos de Servicio Contra el Narcotráfico.	
	Batallón de Acción Directa 'Teniente Coronel Dixon Giuliano Castrillon Gomez'	
Brigada Contra la Minería Ilegal BRCMI 6		Ground forces against illegal mining
Brigada de Fuerzas Especiales Rurales		Special forces
Departamento de Alistamiento para el Combate y la Seguridad Aérea		In charge of the safety and maintenance of the bases

Policía Nacional de Colombia order of battle

	Unit	Equipment
Grupo de Operaciones Aéreas	Compañía Antinarcóticos de Aviación de Bogotá	Basler BT-67, DHC-6 Twin Otter, Cessna 208 Grand Caravan, Cessna 208B Grand Caravan, Beechcraft 1900C, Beechcraft 1900D, ATR 42, DHC-8 Dash 8-311, Beechcraft B200 Super King Air, Beechcraft 300 Super King Air, Beechcraft 350 Super King Air and Fairchild C-26 Metro
	Compañía Antinarcóticos de Aviación de Guaymaral	Piper PA-31 Navajo, UH-60L Black Hawk, UH-60 A, Bell 412EP, Bell 212, Bell Huey II, UH-1H, Bell 206L, Bell 206B-3, Bell 407, Hughes 500D and Cessna TU206G
	Compañía Antinarcóticos de Aviación de Santa Marta	Bell 212

	Unit	Equipment
Grupo de Operaciones Aéreas *(Cont'd)*	Compañía Antinarcóticos de Aviación de Tulúa	Bell Huey II
	San José del Guaviare unit	No assigned aircraft
	Villa Garzón (Putumayo) unit	No assigned aircraft
	Tumaco (Nariño) unit	No assigned aircraft
Grupo de Estandarización Aeronáutica	Escuela de Aviación Policial	Bell 206, Cessna 172 and Cessna 150L
Grupo de Seguridad Integral		Base security
Grupo de Mantenimiento Aeronáutico		Maintenance
Grupo de Abastecimiento Aeronáutico		Logistics

BAAV 3 is equipped with a mixture of Mi-17-1V, Mi-17MD and Mi-17-V5 aircraft for transport.

Glossary

ARC	Armada de la República de Colombia – Colombian Navy
BAAV	Batallón de Aviación
Batallón	Battalion
BAMAV	Batallón de Mantenimiento de Aviación – Aviation Maintenance Battalion
BAEMA	Batallón de Especialistas de Mantenimiento de Aviación – Aviation Maintenance Specialist Battalion
BAOEA	Batallón de Operaciones Especiales de Aviación – Aviation Special Operations Battalion
BAVE	Brigada de Aviación
BETRA	Batallón de Entrenamiento y Reentrenamiento de Aviación – Aviation Training and Retraining Battalion
Brigada	Brigade
CACOM	Comando Aéreo de Combate – Air Combat Command
CAMAN	Comando Aéreo de Mantenimiento – Air Maintenance Command
CATAM	Comando Aéreo de Transporte – Air Transport Command
CIAC	Corporación de la Industria Aeronáutica Colombiana – Colombian Aeronautical Industry Corporation
Ejército	Army
EJC	Ejército Nacional de Colombia – Colombian National Army
Escuadrón	Squadron
Escuela	School
Escuadrilla	Wing
FAC	Fuerza Aérea Colombiana – Colombian Air Force
Fuerza de Tareas	Task Force
FTA	Fuerza de Tareas Aéreas – Air Task Force
GANCA	Grupo Aeronaval del Caribe – Caribbean Aeronaval Group
GANCE	Grupo Aeronaval Central - Central Aeronaval Group
GANPA	Grupo Aeronaval del Pacífico – Pacific Aeronaval Group
Grupo	Group
GROEA	Grupo de Operaciones Especiales Aéreas – Air Special Operations Group
Mantenimiento	Maintenance
Operación	Operation
PAF	Patrullera de Apoyo Fluvial – Riverine Support Patrol Vessel
PNC	Policía Nacional de Colombia – Colombian National Police
Regimiento	Regiment
SATENA	Servicio de Aeronavegación a Territorios Nacionales – National Territories Air Navigation Service

Other books you might like:

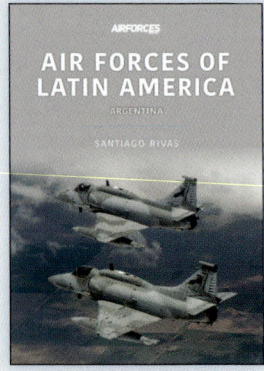
Air Forces Series, Vol. 1

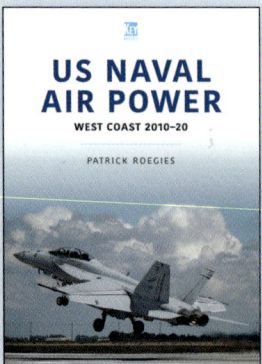
Air Forces Series, Vol. 2

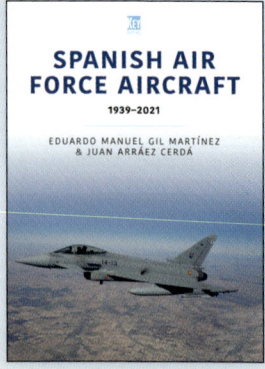
Air Forces Series, Vol. 3

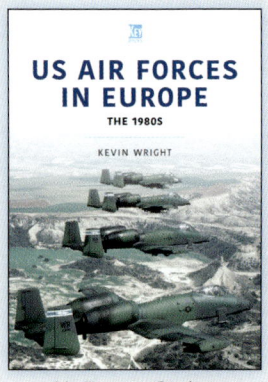
Air Forces Series, Vol. 4

Airlines Series, Vol. 6

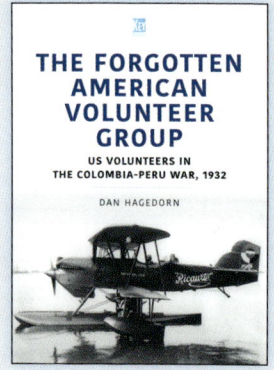

For our full range of titles please visit:
shop.keypublishing.com/books

VIP Book Club

Sign up today and receive
TWO FREE E-BOOKS

Be the first to find out about our forthcoming book releases and receive exclusive offers.

Register now at **keypublishing.com/vip-book-club**

Our VIP Book Club is a 100% spam-free zone, and we will never share your email with anyone else. You can read our full privacy policy at: privacy.keypublishing.com